21st Century Robot

Brian David Johnson

Illustrations by Sandy Winkelman

MAKER MEDIA™
SEBASTOPOL, CA

21ST CENTURY ROBOT
by Brian David Johnson

Revision History for the :

See *http://oreilly.com/catalog/errata.csp?isbn=9781449338213* for release details.

ISBN: 978-1-449-33821-3

Contents

21st Century Robot

A Manifesto

In 21st century technology has progressed to the point where what we build is only constrained by the limits of our imaginations. It's time to imagine a radically different kind of robot. A robot that is designed, constructed and programmed like never before. We can imagine a build a far more amazing future for robots and their relationship to humans. It's time for a 21st Century Robot. This is our Manifesto.

21st CENTURY ROBOT
MANIFESTO

A ROBOT IS:
IMAGINED FIRST
EASY TO BUILD
COMPLETELY OPEN SOURCE
FIERCELY SOCIAL
INTENTIONALLY ITERATIVE
FILLED WITH HUMANITY AND DREAMS
THINKING FOR HER/HIM/ITSELF

Imagined First

Nothing amazing was ever built by humans that wasn't imagined first.

Imagination is the most important skill needed to build your robot. Our science and technology have progressed to the point where what we build is only constrained by the limit of our own imaginations. In the 21st century anyone can imagine and design a robot.

Ask yourself: Who do you want your robot to be? What is your robot's name? (Because—Every robot has a name.) Every robot is an individual. How do you want to act and interact with your robot? What would you want your robot to do that other people could never imagine?

Science fiction stories, comics and movies are a powerful tool to imagine your robot first. We can use science fiction based on science fact to design robots and then share those stories as a technical requirements document.

Easy to Build

Building a robot isn't hard.

Back in the 20th century building a robot was hard and complicated. Computers were massive and slow. Electronics were complicated and the manufacturing process was reserved for just a handful of people that had the money to build factories and assemble lines. But all that's changed.

Today anyone can build just about anything. Computers are small and easily accessible. Software tools and apps allow anyone to be a programmer. 3D printers mean we can manufacture exactly the part we need. There are entire communities, event and places where you can go to build a robot.

Completely Open Source

The idea behind open source is that people should have control over the technology we use. We should be able to build it, modify it and share it. The practice and community around open source really got popular at the end of the 20th century with the internet and software. But since then it has expanded into new areas like electronics, medicine, fashion, government, education and even soda pop. (Open source cola gives away the often tightly held secret of what goes into the beverage)

A 21st Century Robot is completely open source. Starting with the 3D files, everyone should be able to design and customize their own robot. Next the software that runs the robot and makes up its brain is free and open. You can play with the operating system and even design different apps for your robot.

More than anything else we want you to share your designs with others. Did you come up with a cool new leg design? What's your latest app? Even the production of these robots is open, people all over the world can collaborate to build better, smarter, funnier and more exciting robots.

Intentionally Iterative

Why make just one robot when you can make lots of robots?

The practice of iteration is the repeating of a process with the goal of making multiple versions of an object or a project. We make many robots with the aim to improve and experiment with different versions. Each robot is the starting point for the next. Each new robots play around with what we learned from the previous. It's OK to experiment and try new ideas. Building off our open source sharing, things get really interesting when you experiment with other people's ideas

Fiercely Social

A 21st Century Robot is primarily designed to act and interact with people. They are fiercely social, connected to the Internet, social networks and each other. What would your robot say to another robot?

How we build these robots is social as well. There is an entire community of people all around the world who love building and who dream of a very different kind of robot. Working together we can change the future of robotics.

Filled With Humanity & Dreams

Robots are built by people. We design them with our hopes and dreams. We can imagine our possible futures and put those future dreams into our robots.

Japanese roboticist Masahiro Mori saw that our machines, our robots are not separate from us. Because we make robots they are an extension of ourselves. In 1974 he wrote that "Machines, while appearing to be separate from us, are in truth only functions that have been cut away from us, but are essentially part of us."

Technology is just a tool. Humans build tools and we fill those tools with our culture, our ideals and our dreams for the future. Technology is not separate from people; it is an extension of who we are.

Our robots are a way to imagine a different future, to build our dreams and let them play with us.

Thinking For Her/him/itself

You can design your robot to his her/his/its own personality and behaviors. You can make apps to do just about anything your imagination can dream up.

A 21st Century robot isn't a puppet. They are designed to think for themselves. To move around and make decisions. They are designed to act and interact with you and other people. We want them to be adventurous and strange and funny.

Our Motto: "Every Robot has a Name"

Every robot is built by people in their backyards, and garages and basements. Every person has a name. Every one of us is an individual that's why every robot should have a name because your robot and all the other robots you will build will be an extension of you. Your robot and the design for your robot will go out into the world. Other people might use part so of it and then make it their own. This sharing and iteration can go on and on. We keep building...we keep sharing...we keep designing, programming and building robots.

That's a 21st Century Robot. It's a way for us to imagine, design, build and share our own personal visions for the future. And it's also a way to make some really awesome little friends.

every robot has a name

How to Use This Book

The World Maker Faire Edition

This edition of 21st Century Robot is a work in progress. Just like a 21st Century Robot is completely open source so is this book. You can literally see the source code of this edition of book. We are collaboratively building our robots and bringing everyone on the journey with us!

There are four science fiction stories to introduce you the 21st Century Robots and explore how they might act and interact with people in the future. Following each story you'll find a "How to" chapter to show you how to build the robots your reading about in the fiction.

What's special about this edition of the book is that the "How to" chapters are not complete. We did this on purpose! We wanted to include everyone in the development of our robots and even the writing of this book. Over the next few months we'll be collaborating and iterating with robot builders all over the world to bring the 21st Century Robots to life.

But we need your help!

You can be a part of the book. Go to http://www.robots21.com to find out about the Robot Maker Faires and how you can share your designs, ideas and visions for the future of robots.

Welcome to 21st Century Robot—a new kind of book for a new kind of robot!

The Loneliness of the Long Distance Robot

"Was anyone hurt?" Dr. Simon Egerton asked after a long pause. Then...

"Hurt?" Tian Yu, a project manager, asked in a quick baffled shot. Her voice was sharp with precise words that stabbed at the ears. The background noise was so loud she had to yell to be heard. "What do you mean?"

"Was anyone hurt in the accident," Egerton was afraid of the answer.

"I'm sorry Dr. Egerton but you are going to have to repeat that," Yu shouted. I'm on a shuttle now headed to the fab. I can barely hear you."

"Was anyone hurt?" Egerton yelled.

"What do you mean hurt? Who could be hurt? There's no one at Fab 5." she replied. "There was no one anywhere near the place."

"What about the bot in the new HCI test lab, I mean, the Human Computer Interaction validation lab?" Egerton asked. "The one that was just installed. The one I..."

"That's why they wanted me to call you." In the wild noise Yu still seemed bored and annoyed at the same time. "That's where the fire started. The new robot you made...the new bot you designed burned down the entire DeutchConn Fab 5 facility!"

"I'm sorry Ms. Yu but how could the fire start there?" Egerton asked. "There's no fab equipment there. It's just a testing lab. There's nothing there but..."

"They have it on video. They know that's where it started," she answered. "That's why they wanted me to call you. That's why you have to come back to the fab, or well, let's say they started talking about lawyers..."

"But how could a fire have started in the lab?" Egerton ignored the threat; he still didn't understand. "There's nothing there. There's nothing in the HCI lab that can start a fire."

"The techs say the new test bot set itself on fire," she answered. "That's what burned out the entire factory. It was your robot."

"The bot set itself on fire?"

"Yes," There was nothing in Yu's voice but words. The background noise erupted as the cargo shuttle fired its engines to slow down. "There's a shuttle coming to pick you up in four hours. Randall you're picking up the robot guy next right?" she yelled to someone else. "You don't have a choice," she returned to Egerton. "Do you want me to send you the video? You can see that it clearly sets itself..."

"Yes send it," Egerton interrupted and shut off the call.

Cockpit, Cargo Hauler Chen-Ming. 30 Hours to Reboot

"It's a weird little place," the shuttle pilot said as the approached DeutchConn Fab 5. "I take maybe three people a year out here. The shipping drones, hell they run back and forth all the time. I swear nobody knows what they make out here."

"It's just chips," Egerton answered in a low voice. He paused the video of the HCI test bot at the precise moment before it set itself on fire.

"Chips?" the pilot called back over his shoulder. "Like computer chips?" The pilot's name was Randall J. Gan. He was young and liked to talk. His long lean body barely fit in the cramp pilot's chair of the beat up cargo shuttle.

"Yeah, they're just really small and really fast." In the video the robot had doused itself in an unknown liquid and paused before creating an electric spark between itself and the metal table. Lighting itself on fire was no easy task.

"Seems silly putting a chip fab all the way out here." Randall mused to himself.

"They don't need any people to do it," Egerton spoke up before he could stop himself. "They don't want any people." The rest of the big cargo shuttle was empty, no gear, no other passengers just Egerton. Randall has asked Egerton to join him in the cockpit because sitting back there all by his-self was just a little spooky.

"Well who watches them?" Randal turned around as the shuttle neared Deutch-Conn Fab 5. "How do they know that they are doing it right? Building them right?"

"They don't need to know as long as they work. Plus they can watch them remotely. The place is filled with sensors and monitoring systems. People don't need to be there. People are bad for quantum chips." The shuttle was getting close to the Fab 5's dock. The fact that the pilot wasn't watching started to make Egerton nervous.

"But what happens when something breaks?"

"There are bots for that?" Egerton answered.

"There are bots to fix bots?"

"Oh yeah," Egerton finally looked up from the still image of the HCI bot. It had haunted him the entire trip. He stared at every inch of the bot trying to figure out why it set itself on fire. Looking away felt like taking a deep breath of fresh air.

"There are bots to fix bots. Bots to clean the bots. Bots to print out new bots and replacement parts. Even bots to test that what the bots are making is good enough to send to us human."

"Really?" The pilot stuck out his lower lip in disbelief.

"Shouldn't you?" Egerton pointed to the heads up display that was now filled with the FAB's docking station.

"Nah," he pilot waved away Egerton's concern. "There are bots for that as well." The pilot's laugh was deep and wet laugh. "I'm just here in case something goes really really wrong."

Egerton smiled and retuned his eyes to the HCI bots on the screen. He let the video play and watched the bot set itself on fire.

"Docking in 60 seconds," the warm female voiced computer said.

"There's my Shelia," the pilot tapped the heads up display. "I was getting worried you were mad and had left me."

"I couldn't leave you Randall," the warm voice replied. "I love you too much."

"That's my girl," Randall ran his fingers gingerly across the screen.

"Docking in 30 seconds," the warm voice reported.

"Dr. Egerton," Randall's voice grew serious.

"Yes?" the tone forced Egerton turn from the burning bot.

"Over the last 12 hours, I've hauled in construction dudes, fast track builders and even a whole rig full of hopped up code monkeys. Not to mention that crazy woman Yu...But you're here alone on this big old rig and I got this priority order to come pick you up. Can I ask what you do?"

"I guess I'm like you," Egerton smirked. "I'm the one they call when things go really really wrong."

DeutchConn Fab 5, Main Floor. 25 Hours to Reboot

"Did you see there aren't any beds? I mean I don't have a crew of that many people but at least they could have given us beds...," Yu complained as she led Egerton through the charred remains of the production fab. "This is where the fire was the worst," she continued. "We're going to do this fast. Right now all my crews are working on the quick work. I've got a team of 5 guys and a mess of our own recovery, construction and security bots. We want to show progress. Deutch likes to see progress."

"Is that..." Egerton stopped and looked up at the towering security bot that guarded the door to Fab 5's main floor.

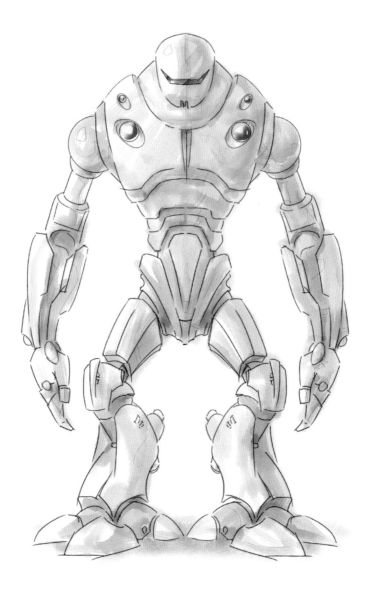

"Huh?" Yu stopped when she realized Egerton had stopped and wasn't following her.

"Where did you get this?" Egerton asked taking a hesitant step towards the hulking machine. It acknowledged him but didn't move.

"What do you mean where did I get it?" Yu was growing impatient. "We've had that piece of junk since it was new. It's only good for..."

"But this is the Gen I version," Egerton looked at the massive arms and thick legs. "I haven't seen one without its riot pads in years."

The bot shifted its broad feet and Egerton jumped back in fear.

Yu laughed. "Come on Dr. Egerton. Stop staring at my old bot. No one cares about some old security bot all the way out here. Even if he could take you head off with a jab of his elbow."

"But he's dangerous," Egerton caught up with Yu and they continued.

"He's only dangerous is someone messes with him," Yu replied. "That's what makes him a good security bot. We can't let anyone get in the way of our schedule."

"It's an aggressive schedule," Egerton answered. He really didn't know what to say. He didn't know much about the repair plan. He was still spooked by the sight of the security bot. It was like seeing a hulking wild animal sitting quietly in your neighbor's living room.

"Yep. That's why no beds," Yu stopped and turned to Egerton. "We have 25 hours to get this place repaired and back online. 25 hours...There are no beds because they don't want us to sleep."

"It is aggressive..."

"It's insane but I can do it," Yu smiled proudly. "I've seen worse than this. Fires aren't common in Fabs or hell even in any stations. Why would anything catch on fire..." she scanned the soon.

The main floor of DeutchConn Fab 5 was a wide flat room with a high ceiling that gave room for the bots to work. The entire room sat on an elaborate system of jacks and suspension monitors to keep it eerily still not matter what happened. The arms and tracks of the production like had been scrapped clear from the floor and where piled in a charred pile in the back corner. The sight of the destroyed robots disturbed Egerton. They looked like victims of an atrocious war crime.

"We had one job where the fire nearly breached the outer hull of the station. Literally it was inches between us and a heck of a lot of trouble. None of the guys would make the weld and put in the repair plate. I had to do it myself." Yu shook her head. "All these big quick construction dudes and they're afraid of a little crushing vacuum from space. SLRRRRRRP." She sucked in her cheeks and slapped her hands together. Laughing she continued to lead Egerton to the HCI lab.

"They're a bunch of babies."

"When do the new fab bots arrive?" Egerton asked.

"The arms and tracks?" Yu asked?

"Yeah the new ones?" Egerton pointed to the dead pile of bots.

"Oh them, I guess I never thought of them as robots," Yu answered flatly.

"They are..." Egerton waited. "The industrial bots, when do they arrive?"

The air got tense.

"They arrive 19 hours before we go online," Yu wasn't sure what she has said to upset Egerton.

"The HCI Lab is this way right?" Egerton pointed to the far door.

"Yes."

"Is the HCI bot still in there?" Egerton moved towards the door.

"Yes."

"Do you need anything else from me?" Egerton asked Yu. "I think I'm going to have my hands full for the next 24 hours."

"Nope," Yu replied. She looked at Egerton like he was a little too strange for her taste.

"Please come get me when the new bots arrive," Egerton's voice softened with worry as he looked around the carnage of the room.

"Are you OK Dr. Egerton?" Yu asked. "Are you worried about something?"

"I'm worried that I won't figure out what happened in that room," Egerton pointed to the HCI Lab. "And all of this..." he gestured to the fire damage and destroyed bots. "...and all of this will happen again."

DeutchConn Fab 5, HCI Lab. 22 Hours to Reboot

"It looks like an autopsy room in here," a compact Asian man said leaning against the door.

"It is," Egerton replied without looking up from the table.

The HCI bot lay on the table. Egerton has spent that last three hours gingerly dismantling what was left of the bot. It was a mess.

"My name is Shanwei," the man at the door said casually. "I'm running the sub-crew for Yu."

"Ok..." Egerton didn't glance up. He didn't care. He was deep in thought. He picked away at the corpse of the bot. Every piece of the bot had fused together with itself. Its outer body had melted into the servo motors. The battery and the brain were locked together in a thick and dangerous blob.

"They say that you built him," Shanwei spoke after a long silence.

"I did," Egerton tried to pull away the bot's drive rods from its melted inner body. By design the HCI bot had not legs. It had a torso, arms and a head. It was fixed to the end of the fab process to test a statically valid number of chips as they came off out of the fab.

"Do you know what happened?" Shanwei asked. "Yu's worried you're going to burn the place down again. Not that she cares. It's just means we all get paid to do this again..."

"I think it killed itself." Egerton snapped the chest cavity to free the brain.

"What?" Shanwei laughed and was sure he didn't hear the intense doctor correctly. "Did you just..."

"Yes," Egerton cut in. "I think that this little guy killed himself...committed suicide...and no I don't know why..."

"Okay...okay," Shanwei stepped back away from the door. "We're all just curious. I mean..."

"It had to have wanted to do this to itself," Egerton wasn't listening to Shanwei. "Why would he do this? It's like he wanted to melt everything..."

Realizing Egerton was in his own world; Shanwei turned and left the lab. Egerton never looked up from the bot.

DeutchConn Fab 5, Load Dock. 19 hours to Reboot

"Have you figured out why the bot committed suicide?" Yu asked as they watched the shuttle carrying the new robots dock with the station.

"How do you know the bot committed suicide?" Egerton asked.

"One of my subs told me," Yu replied. "He says that you said the bot killed itself. That can't be true."

"It is..." Egerton was amazed at the delicate ballet between the massive shuttle and the even more massive station. He saw them as two gigantic robots slowly dancing into an embrace.

Once the shuttle was safely locked in Yu asked, "So what are you going to do?"

"Build another one," Egerton said moving to the opening cargo doors as they revealed a stunning landscape of gleaming robots.

DeutchConn Fab 5, HCI Lab. 10 Hours to Reboot

Egerton ran his finger slowly underneath the bot's neck and gently turned it on. The fluttering and familiar sound of the bot coming to life relaxed Egerton's mind.

"You're running out of time," Yu said knocking on the door.

"What do you mean?" Egerton looked up from the little bot. The new bot was different than the original HCI bot. He had the same arms, head and torso but this time Egerton had printed some legs for him.

"Can I show you something?" Yu gestured over her shoulder.

"Can it wait? I'm bringing this test new bot online. I need it to look at the A.I. before I finish printing the new HCI bot and I want to..."

"No it can't wait," Yu snapped. "We have 10 hours before we go online and from what I can tell you are no closer to figuring out what happened. Frankly I'm worried that this whole place is going to..."

"Ok, ok," Egerton moved over to the door. "Don't worry. This new bot is based on the same architecture from the one before. I'm going to bring him up and run through some tests I think I have an idea what happened." Egerton stopped at the door, leaving the bot alone in the room.

"This is your last chance," Yu gestured over her shoulder. "We're almost done."

"Wow!" Egerton was amazed at the transformation. The Main Fab Floor was transformed back to its pristine state. Walking into the room he could see that only a few contracts remained in their robo-assist gear doing the detailed finishing work at a blistering pace.

"I know," Yu smiled, pleased with her work and her crew. "This isn't our first rodeo. But this is your last chance. We need to get the new bots installed and then clean this place out. Humans are hairy and bad for chips. It's going to take a few hours to clean this place down to spec."

"How long do I have?" Egerton grew worried. "I need more time."

"I can give you an hour," Yu replied. "Hour and a half tops but then I need to clear it and sanitize it. No people in here ever again."

"Ok, I'll do my best but..." Egerton paused. The little bot was standing in the door of the HCI lab watching Egerton and Yu.

"He's cute," Yu said.

"I didn't think he could do that," Egerton was both amazed and a little scared.

Yu walked over to the little bot and the little bot looked up at her. "What's his name?"

"I don't know," Egerton answered.

"How about Jimmy," Yu said. "I once had a gold fish with big ole eyes like that. His name was Jimmy. Yeah...he looks like a Jimmy."

Egerton stood next to Yu and looked down at Jimmy. The little bot looked back.

"You have no idea what happened do you Dr. Egerton?" Yu asked flatly. "You have no idea why that bot killed itself do you?"

"No," Egerton answered. "I have no idea."

DeutchConn Fab 5, Load Dock. 5 Hours to Reboot

Jimmy looked small and out of place in the massive loading dock. Randall and the Chen-Ming had returned from a quick trip ferrying the subcontractors that had finished their work away from the fab.

"Yo Doc, who's your little friend?" Randall asked sliding down the cargo shoot.

"His name is Jimmy," Egerton replied. "Jimmy this is Randall, he's our pilot. Say hello."

"Hello Randall," Jimmy said walking over to the pilot.

"Would you look at that! He's a cute little fella," Randall patted the bot on the head. "What's he for?"

"Jimmy's a test bot. He has the base AI from new HCI robot," Egerton explained. "You see, Jimmy is different than all the other bots in the Fab. He's designed to interact with people. He's social. All of those other bots are designed to work alone or with other bots but Jimmy and the HCI bot are different. We use the

HCI bot to validate the human experience on the chips. Could it perform to the spec? Did it recognize some base emotions stuff like that..."

"I'm very close to not following you anymore," Randall confessed.

"Nah, it's no big deal," Egerton continued. "Jimmy and the HCI bot are just quality testers. These new quantum hybrid chips are so fast and powerful that they are meant to understand the person it's working with. I built these bots to be more human, they are more complex so they can test the chips."

"Makes sense," Randall nodded.

"We're locking the place up," Yu said as a small army of human and bot contractors hauled their gear into the Cheg-Ming. "I need to run the whole fab dry for the next two hours. That means all these knuckleheads can get out of here. Take'em away Randall..." She waved away most of her team. "Looks like it's just you and me and a few techs and a whole lot of robots." She pointed through one of the few observation windows into the Fab. "Strike it up!" Yu yelled and the bots hesitantly came to life.

"Good job Yu," Randal said climbing back to the Chen-Ming. "I'll drop your boys off and then be back here for you and the doc and his little bot."

"Keep'em safe," Yu yelled.

Egerton marveled at the delicate and precise beauty of the Fab 5. The arms raced about, spinning on multiple axis while the movers trundled dutifully along. "We'll keep her dry and test the loads," Yu said again. "I want to start doping the silicon in two hours then we can start the Fab running full steam. That's when we'll need your HCI bot up and running."

"She's running now," Egerton showed Yu a feed from the HCI lab. The bot with an upper body like Jimmy sat still and lifeless by itself at the end of the production line.

"What's his name?" Yu asked.

"Her," Egerton smiled. "I named her Little Ling."

"Cute," Yu smirked. "But is she going to burn this place down?"

"I don't know yet." Egerton replied.

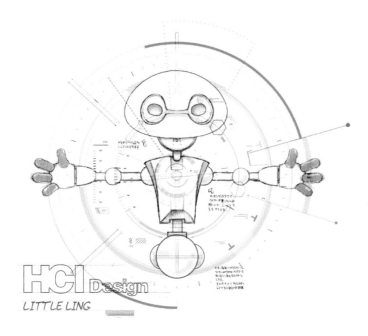

LITTLE LING

Cargo Hauler Cheng-Ming. 1 Hour to Reboot

"Look at those little guys go," Randall said as the Chen-Ming pushed away from Fab 5. Yu had triggered the cleaning bot to scrub down the loading dock for any stray contaminants. The little bots swarmed and cleaned the massive dock in swooping clocks of tidiness.

"This is good," Yu said and Randall stopped the Chen-Ming. "Give me some eyes," Yu called back to the few remaining techs. The front quarter of the cargo hold had been transformed into a make-shift command station. From this distance they could monitor the health of the Feb 5 to make sure it was back in order. The screens were alive with video feeds and data.

"Let's see how she's doing," Yu watched the data.

Egerton watched the single feed coming from the HCI lab. Jimmy stood by his side, careful not to get stepped on. The bot looked a little nervous in the roar and commotion of the shuttle so he stuck close to Egerton's leg.

"Ok we're locked and steady," Randall popped out his chair and came over to watch the show. "I told them we'd bug out of here in 45 minutes or less."

"Right on schedule," Yu said to herself.

Egerton glanced down at Jimmy then returned his tense gaze to Little Ling's video feed.

"You bot ready?" Yu asked, motioning to the video feed.

"She's been online for the past 35 minutes," Egerton answered.

"How can you tell she's on?" Randall asked. "She's not moving."

"I turned her on," Egerton said. "She's just waiting for work to do."

"How do you know she won't destroy all my work?" Yu asked. "I'm kind of proud of what we did here. 25 hours from a charred wreck to a working fab is pretty amazing. So Dr. Egerton how do you know she's not going to burn it down?"

"We don't," Egerton replied calmly.

"What?" Yu was stunned.

"With these new bots you can't know for sure," Egerton explained. "Just like you can't always know for sure what a human is going to do. You just need to treat them right and wait and see."

The first chip came into the HCI lab for testing and validation. Little Ling slid it into the test screen and began interacting with it. Her test was swift and accurate. It passed and she sent it to shipping.

"But that's insane!" Yu got very close to Egerton. "We can't spend all that time and money and just hope it will all be ok."

"Yes we can," Egerton smiled. "It just like Randall here..."

"What?" Yu was getting impatient.

Little Ling started testing another chip.

"Wait now I'm a bot?" Randall smiled.

"It's about trust," Egerton said. "You trust that Randall will park the Chen-Ming where he needs to park it. And when we are done, you trust that he'll take us out of here. You don't know if he's actually doing to do it. He could hit the thrusters and ram is straight in the Fab's dock..."

"Don't be stupid," Yu spat.

"It's true," Egerton continued. "You have a reasonable trust that Randal isn't going to wreck the Chen-Ming until he does something to counter that trust then you'd react. It's the same thing with Little Ling. She's seems fine now." Egerton pointed to the still and quiet bot as she waited.

"But she did burn the place down!" Yu looked at the data streams nervously and then looked back. "I mean I'm not the crazy one right? We were all here, she did burn it down."

"That wasn't Little Ling," Egerton spoke up. "That was a very different bot."

"But..."

"I know why the other bot killed itself," Egerton interrupted.

"Why?" Yu insisted.

"Because it was lonely" Egerton answered.

//// Need another time-stamp here maybe? I like breaking it off at Egerton's conclusion above.

"How can a robot be lonely?" Yu snapped.

"These are no regular bots," Egerton said looking down at Jimmy. "They're designed to interact with people. Their brains are kind of like our brains." "I'm not so sure I want a robot with a brain like mine," Randall chuckled.

"That's what make's Little Ling so good at testing how people will use the chips," Egerton watched as the bot came to life and did her job, then went still. "You see her brain is made up of personas. But only a handful like humans."

"Are you saying we all have multiple personalities?" Yu asked.

"No," Egerton smiled. "We all have different modes of thought. Sometimes we're working, sometimes we're learning, sometimes we're talking with friends. But I realized that with the HCI bot all the different companies that were testing their chips were filling the bot with their own audience that they wanted their product tested for. Some were targeting middle aged mothers while other's wanted to sell to teenagers, so they just uploaded that person to be tested."

"How many people did the bot have in it?" Randall asked.

"Over two hundred."

"Yikes," Randall watched Little Ling.

"So, I limited the number of people that could be uploaded," Egerton added.

"But why did the bot destroy itself," Yu asked with sincere interest.

"Well...way down deep in that HCI bot was the original learner persona I built. But it never got used. It never learned anything because people kept cramming new people into its head. Finally it must have gotten threatened or stressed or...."

"Lonely?" Yu said.

"I think so," Egerton replied. "Then it was just a matter of time before the original personality saw all the other personalities crammed inside as threats or the source of its loneliness so it protected itself by destroying them."

"...and destroying itself at the same time," Yu nodded.

"Yes, I think so."

"That's kind of sad," Randall spoke up.

"So how do we know it won't happen again?" Yu asked. And don't say we can't know. You had to do something. How do we make sure that Little Ling here doesn't get lonely?"

"I gave her friends," Egerton smiled and pointed at Jimmy.

The trio was quiet for a moment as they watched the little bot. He stuck close to Egerton's leg.

"But he's here and she's there," Yu pointed at the screen.

"That's easy," smiled again. "I gave them a little network so they can talk...so they can be social. It's open so people can play with it. It's a little network just for robots."

Just then the loading dock doors of Fab 5 opened and a cargo drone slipped out.

"First shipment verified," the lead tech said without emotion.

Yu, Randall and Egerton watched the slim ship slide by and make its way to the transport hub for distribution.

"Fab 5 back online," Yu reported to the Transport hub. "We are back online. Everything looks good. We're pushing out of here."

"A network for robots?" Randal looked down at Jimmy. "Are they talking now?"

"Yep," Egerton watched Little Ling sitting still waiting for something to do.

"So Jimmy tell me," Randall said. "How's Little Ling doing? Is she lonely? How's our girl?"

Jimmy looked up at the tall man and replied, "She's OK. She's a little bored but she's OK."

Design the Body

Sneak Peek! This is a summary of the chapter to come. Below is the journey we are taking into the future of robots.

A 21st Century Robot is imagined first. It is designed in your imagination. Our robots started out that way nearly ten years ago in a collaboration between myself and roboticist by the name of Dr. Simon Egerton (yes he's the same Dr. Simon Egerton that's been fictionalized in the science fiction stories). We start off our Design the Body chapter by chatting with Simon and the rest of the Creative Science Foundation to discover the scientific origins of Jimmy, our little robot.

Next we meet our illustrator Sandy Winkelman. Sandy has been on this journey with Simon and me the entire time, visualizing and collaborating with us on the appearance and personality of the robots. In the chapter Sandy tells us how he first envisioned Jimmy and what inspired him.

To move Jimmy from Sandy's two dimensional illustrations and into the world of 3D we needed Wayne Losey. Wayne is a toymaker and 3D designer. He tells us about his process of bringing Jimmy from our imaginations into the physical world. You'll get access to Wayne's 3D design files so you can play and modify Jimmy. Or if you have your own idea you can design your own 21st Century Robot

To get Jimmy moving we head to Olin College and meet David Barrett. Dave, along with his class of innovators and builders, are on a mad dash to design and build Jimmy's body. But they aren't just building one Jimmy, they're building ten! Like our Manifesto says, a 21st Century Robot is intentionally iterative and in that spirit the Olin crew is exploring multiple versions of Jimmy so that we can all collaborate, iterate and share as many ideas and versions of Jimmy as possible.

This all leads up to November 2013. Dave and the Olin crew will unveil their reference designs to a collection of students, makers and robot enthusiasts for our first 21st Century Robot Maker Faire. Next we'll go to Los Angeles and other city as well.

You can be a part of designing and building 21st Century Robots. You can even be a part of this book! Check out *http://www.robots21.com*, come to the Maker Faires, join on line and add your inspirations and designs to the project.

I AM Robot

Grande Lobby - Villa Marquis Station

"Dr. Egerton we are so pleased you could make it here so quickly," Sergio Sauer extended his perfectly manicured hand. "This is rather a delicate matter that we... well I'm sure you know."

Sauer's hand was warm and soft. Egerton shook it quickly and replied, "Shanwei explained what happened but I don't understand...I know you want us to look at your elder car bots but I'm not really sure what..."

"Please Dr. Egerton not here... *here* is a very bad place to talk. Let us move upstairs." Sauer ushered Egerton through the grand and expansive lobby of the Villa Marquis. The upscale retirement community exuded the carefully blended feeling of a resort hotel with a bustling urban high rise. The Villa Marquis was one of the more desirable communities in the clog of stations that ringed Earth. "The lifts are this way..."

Egerton watched the elevator pods as they shot up through the lobby. The cluster of pods lay at the exact center of the leather lounge chair and murmuring Zen fountain filled lobby. "Wow," Egerton said. He had not seen elevators like this in decades.

"I know they are quite impressive," Sauer said with a self-satisfied grin.

"Actually I was thinking how inefficient it is to have elevators like that in a space station," Egerton replied. "It's a lot of wasted space."

"At the Marquis we have found that our residents prefer the older style," Sauer wasn't pleased. "Everything we do here is for the pleasure of our guests." The way Sergio Sauer said the word pleasure made you think about warm salted caramel or a bath tub brimming with heavily scented bath salts.

"I can see..."

"This way Dr. Egerton," Sauer ushered him into the private service elevator and typed in a code. The pod shot up 30 stories through the lobby of the Marquis and disappeared into the heart of the space station.

"Can we talk here?" Egerton asked.

"No," Sauer pointed to the surveillance sensor.

"But this is your property why should you be worried about..." Egerton started.

"It's not me I'm worried about," Sauer replied. "It's your robot that nearly killed one of our guests, Dr. Egerton. I'm worried about you incriminating yourself."

Edelman Guest Residence 2145 - Villa Marquis Station

"Sergio tell them to let me put my legs on!" the large woman said from the mini sofa when Sauer and Egerton entered the small apartment. Two polite but stern looking nurses stood off to the side of the apartment tapping at their screens.

"Now Ms. Edelman," Sauer glided to her side. "We need to keep everyone out of your bedroom for just a little while longer. We went over this remember? This is Dr. Egerton and he made your robot..."

"You made my Cutie?" the woman's old eyes widened. She was quite a large woman but it was plain to see that both of her legs had been amputated just above the knee.

"No, I'm sorry, I didn't design the Cutie that's another company that..." Egerton tried to explain but was cut off.

"He designed its brain," Sauer took Ms. Edelman's hand gingerly.

"Well not really its brain either. That was..." Egerton was cut off again.

"Ms. Edelman is one of our longest and most loved guests here at the Marquis," Sauer pointed at Egerton. "And this is Dr. Simon Egerton; he needs to inspect your Cutie to find out what happened."

"I'm not the oldest," Ms. Edelman glanced sideways at Sauer with dramatic flourish. "I'm just 102. There are plenty of old people here. I've just lived here almost the longest, starting back when the diabetes took my legs." She slapped her right thigh with her right hand. The hand was discretely bandaged. "Can you get my legs for me Dr. Egerton? They are just in there next to my Cutie."

"This shouldn't take long," Egerton smiled and started the three short steps to the bedroom.

"I just want my legs," she shot back. "You can tell me later why my Cutie attacked and stabbed me."

***

The scene in the small room brought Egerton to a dead stop.

"It's horrible isn't it," Sauer whispered and he pushed Egerton deeper into the room so he could close the door.

"What happened exactly?" Egerton asked as his eyes flicked between the blood stained bed, the two pink robo-assist legs and the destroyed robot.

"We found out about it when Ms. Edelman hit her panic button," Sauer pointed to the side of the bed. "All our guest rooms have panic button. Press it and some will be to your room in less than 30 seconds. It's something we guarantee here at the Marquis. We are known for it. You have no idea how expensive it is to..." There was blood all over the wall and the panic button.

"The robot stabbed her right hand?" Egerton asked. "I saw the bandage."

"Shanwei said you were observant," Sauer rolled his eyes. Shanwei was Egerton's partner. The two had started working together after their success at the DeutchConn Fab 5. Shanwei had approached Egerton and with each investigation their reputation had grown. "Look Dr. Egerton this incident is costing us tens of thousands of dollars a day. We need to continue our Cutie roll out if we want to get the health care discounts agreed to by our elder care providers. That one over there," Sauer pointed to the smashed robot. "That one was only number three. We have over two thousand other Cuties to deploy in each of our guests room. The savings will be astro..."

"I saw the boxes of Cuties when we came in," Egerton remembered the strange sight of the little robots in their transparent boxes standing silent outside each door. They looked like motionless adorable three foot guards waiting to be set free.

"We thought that would be a nice touch," Sauer explained. "The guests don't like change. Having the Cuties around and waiting has made the transition easier. Of course the guests would all rather stay with the existing nursing staff but we can't afford that....so the Cuties it is." Sauer sighed.

"A Cutie in every room," Egerton said to himself as he got on his knees over the destroyed robot. Cuties were a popular brand of personal robots. They were similar to the design of Egerton's test bot Jimmy but mass produced and purposely cuter. Shanwei was sure that Egerton could have sued Spinnaker Systems, the maker of the robots, for stealing Jimmy's design. But Egerton knew he'd never win the fight. Spinnaker made its name and money on military bots and big-class drones. The Cuties line of robots were their attempt to get into the consumer market. But it was Spinnaker's use or mis-use of Egerton's open-source robot network that had brought him to the Villa Marquis.

"When the nurses arrived, the Cutie was on the bed standing over Ms. Edelman holding that letter opener," Sauer pointed to the blood covered stainless steel instrument.

"And they destroyed the robot..." Egerton couldn't mask the contempt in his voice. "Did they have to destroy the bot?"

"Look at all this blood," Sauer spat then brought his voice down to a whisper. "I think they did the right thing. Your robot attacked an elderly legless woman while she slept."

"It's not my robot," Egerton searched through smashed pieces of the bots outer and inner skeletons. "I have nothing to do with Spinnaker. They used my open source network to get your bots to work together. They took my network...I had nothing to do with this..."

"That's not what the media is saying," Sauer spat back.

"I'm aware of what the media is saying..." Egerton remained calm. Shanwei had coached him to remain calm. "...and what they are saying isn't true." Finding the Cuties small brain Egerton stood and turned to Sauer.

"That's not what the authorities are saying either Dr. Egerton," by this point Sergio Sauer had lost all his charm and revealed himself to be what he truly was, a mean little administrator of a high end retirement home.

"You know that's not true," Egerton stopped and picked up Ms. Edelman's pink robo-assist legs. They were lighter than he had expected, very high end. "That's why the Marquis is paying me to find out what happened inside here." Egerton took out the Cuties' brain and showed it to Sauer.

"And when will that be?"

"I don't know," Egerton answered.

"What do you mean you don't know?" Sauer snapped. "Every day we don't bring those Cuties online we're losing money."

"Look Mr. Sauer," Egerton slipped the brain back into his pocket. "How about we give the old lady her legs back and you let me go to work? That's the only way any of us are going to find out." Egerton opened the door and walked out.

***

"I've got your legs Ms. Edelman," Egerton said as he returned to the living room. "They are really nice."

"I know," Ms. Edelman took the legs and quickly slipped them on. They charged up and she stood. "My great grandson is a very successful in the stock market. They were a gift." Ms. Edelman was now taller than Egerton. "What happened to my Cutie?"

"I'm not sure," Egerton answered as Sauer slipped back into the room. "But I should know soon."

"Please hurry," Ms. Edelman seemed much calmer and less elderly with her hot pink legs carrying her around the small apartment. "I liked the little guy. He was a big help with my medications...I named him Maury after my departed husband. He always..."

Egerton smiled. "That's a good name..." the sound of the incoming call stopped him. It was Shanwei. Egerton looked into the screen and heard the unmistakable sound of children screaming in terror.

"Simon...Simon can you hear me?" Shanwei yelled. "Simon you have to get here now...it happened again!"

Transition Tunnel and Shark Tank - The Five Worlds Resort

"Please tell me you know what's going on," Shanwei couldn't look at Egerton. He kept his eyes forward as they moved from Pirate World to Candy World.

"Slow down," Egerton grabbed Shanwei's arm. "Jimmy can't keep up." The little robot was trying his best to keep up the pace but couldn't.

"Tell me you know what's going on," Shanwei stooped down and picked up Jimmy.

"He doesn't like to be carried..." Egerton said.

"I don't like..." Jimmy tried.

"Tell me you know what's happened," Shanwei was serious this time. "You haven't seen what I've seen. Wait til you see what's going on down there." He pointed down the tunnel that connected two of the five different worlds offered by the Five Worlds Resort. It was a popular family vacation spot all year round and was conveniently located near three of the largest shipping stations. It was also very close to the Villa Marquis.

"I'm sorry Shanwei," Egerton replied. "I don't know..."

"You always say that!" Shanwei was tense. "You always say you don't know but you do know. You always know...every time these last few months you know but you say you don't know." Under Shanwei's arm Jimmy watched the sharks swimming all around them. The transition tube ran inside a massive shark tank filled with Tiger sharks, Hammerheads, Great Whites and a few gen-mod sharks that were over 20 feet long.

"But I don't know," Egerton answered. "How could I know yet? I just got back from the Marquis..."

"But you built it...you build the network," Shanwei watched a big gen-mod shark swim by and cast a heavy shadow. "You have to know..."

"I'm sure I can figure it out," Egerton tried to calm Shanwei down. "I brought Jimmy because he can talk to them. He can access the network and maybe..."

"Do you really think it's..." Shanwei started and stopped.

"What?"

"No, it's stupid," Shanwei shook his head. "But you haven't seen what's down there...it's....it's really bad." Shanwei continued walking.

They were silent for a while as they passed through the sharks. The sounds of chaos started to scratch its way up into the tunnel. "Do I think it's what?" Egerton pushed.

They reached the entrance to Candy World. It was a brightly lit wonderland of peppermint sidewalks and lollypop street lamps. By the SodaPop Pools a group of fifty children all under the age of eight stood screaming. They clutched their Candy-Colored Cuties to their chests and fought as their parents tried to take them away. A rugged gray emergency vehicle looked garishly out of place as it attempted to fish something out of the pink water of the pool.

"Do you really think the robots are trying to kill us all?" Shanwei asked.

SodaPop Pool – Candy World - The Five Worlds Resort

The screaming had died down to sporadic whimpers and whines.

"Spinnaker wanted to throw a promotional party for their new line of bots for young kids. These Candy-Colored Cuties are pretty like the original Cutie bots..." Danny Tepper explained. She was an athletic woman with the kind of fresh happy face that kids loved. She wore a pink t-shirt that read, "Candy-Colored Cuties are the BEST!". All the adults in Sweets World except Shanwei, Simon and the rescue team were wearing the bright shirts. "...but they gave them that candy look and taught them a few songs..."

"What happened?" Egerton asked.

"The cameras are coming," another adult informed Danny. "What do we do Danny? The cameras will be here in five minutes. They are coming from dock four."

"Just relax Shanni," Danny assured her coworker. "Get the kids and their parents ready to go back to their rooms." She tunred to Egerton. "Dr. Egerton it sounds like you have five minutes before you start losing witnesses."

"Witnesses to what?" Egerton asked.

"Shanni go ahead and get them ready," Danny urged. "Get all the families ready to go. We don't want them here when..."

"We should talk to the little boy," Shanwei set Jimmy down next to him and both watched as the rescue crew began to pull something out of the pool's pink water.

"Yes, he's over here," Danny led the way.

The squat gray rescue vehicle had deployed its arm out over the water. A diver had hooked a cable to something under the water and how the team was drawing it out of the water.

"Ms. Webb," Danny stopped by a woman wrapped in a blanket. She was leaning against the back of the emergency vehicle.

"Yes..." Ms. Webb looked dazed and disoriented.

"Ms. Webb," Danny continued. "We're going to take you back to your room right now. But before you go do you think it would be alright if Dr. Egerton here spoke to Kyle?"

"Is this the man they are all talking about?" Ms. Webb looked at Shanwei. "Did you build those things...those things that tried to kill my baby?"

Shanwei froze.

"I think you mean me," Egerton stepped up. "I didn't build these robots but I think I can help."

"I hope you go to hell," she seethed.

"Ms. Webb would it be alright if I took Dr. Egerton...I think he can help. I'll be there the entire time...if anything..." Danny tried again.

"Just go! But don't you let that's robot near my Kyle" she screamed and pointed at Jimmy. Jimmy ran in the opposite direction of Ms. Webb but Shanwei caught him before he got too far away.

"My supervisor told me you guys could tell us what happened," Danny said confidentially to Shanwei. "You can tell them that it's not our fault."

Shanwei handed Jimmy to Egerton and replied, "We can help. We just need to talk to the boy and then get the..."

"Kyle what are you doing?" Danny asked. "Where's your daddy?" The boy stood on the edge of the pool and stared into the lapping pink water. He was small and pale with thin little bones and half dried hair.

Kyle didn't respond.

"Can someone find Mr. Webb?" Danny called to the other pink-shirted Candy World workers.

"Hey buddy," Shanwei squatted down next to the boy. "You wanna get away from this pool? I want to introduce you to someone."

Kyle looked at Shanwei and then back to the water.

"Is this Kyle?" Egerton asked. He lagged behind because he had put Jimmy down and let him walk on his own.

Kyle saw Jimmy and asked, "That yo robot?"

"Yes, his name is Jimmy." Egerton pushed Jimmy to Kyle. "Say hello Jimmy."

"Hello," Jimmy was about the same size as the boy.

"Hi Jimmy," Kyle didn't seem interested. He stared at the water. It had gotten more choppy as the crew's rescue arm drew in the cable, hauling up something bright and blue.

"Kyle," Egerton started. "What happened? What happened with *your* robot?"

"Kyle's Kady-Co Cootie...push me in," Kyle said to the water. "I can swim."

"He can't swim," Danny translated. "His mother jumped in and grabbed him. But then..."

"What happened after your mom got you out of the pool?" Egerton asked the boy.

"Kyle's Kady-Co Cootie push me yin again!" the boy squeaked. "He hole me down un-the water...Grab my neck..."

When the bright blue Candy-Colored Cutie broke the surface of the pink pool it started to swing on the cable. Initially the robot was frozen and glistening. Everyone around the pool stopped and stared. Then the little blue bot began to writhe and thrash on the cable with a wild animal frenzy. It tore at the cable trying to free itself, trying to get to Kyle.

Kyle began to scream a high pitched scream.

Jimmy ran in the opposite direction of the screaming boy as fast as he could.

What other children were left started to scream as well.

Off in the distance of Candy World the cameras and media showed up.

"Oh no," Danny stood and gasped.

Fairy Hotel – Princess World – Five Worlds Resort

"Even the toilet has sparkles in it," Shanwei said coming from the bathroom. "They are serious about their fairies here." All of Five Worlds Resort became a media frenzy. The only room that Resort management could free up was in the Fairy Hotel. While Egerton worked, Shanwei inspected the room that all three had to share. Jimmy sat in a little chair that seemed made for him, except for the magical stars and butterflies that hovered over his head.

"Hey listen," Shanwei said after finishing his inspection of the enchanted room. "I'm sorry about what happened back there."

"What?" Egerton wasn't paying attention. He was busily working on Ms. Edelman's robot Maury and the Blue Candy-Colored Cutie that had fought to kill Kyle.

"I'm sorry I freaked out back there in the Candy World," Shanwei continued. "It's just...it's just you know all those kids got to me. It was pretty rough."

"We've seen worse," Egerton didn't look up. They had seen worse. The pair had been contracting together for some time and had gained quite a reputation for success when dealing with a particular kind of problem. If you had a problem with your bots, something that no one could figure out they were the team for the job.

"Yeah, it's just the kids..." Shanwei tried to explain. "It's kids...it's a thing with me. I don't like it when kids are in trouble...It's just..." he stopped there.

"Good to know," Egerton moved on. "There's nothing wrong with either brain. They are both fine."

"So that means..."

"That means it's the network not the individual bots brain," Egerton explained. "These Cutie brains are pretty simple but most AI is based on a three tiered architecture. Jimmy is the same way. And there's nothing *wrong* with the brains."

"So now what?" Shanwei asked then jumped in "...and don't say I don't know!"

The enchanted room was quiet. Stars and butterflies hovered above Jimmy's head. Finally Egerton said with a smile, "I'm not sure."

***

The angry swarm of media that had infested the Five Worlds Resort meant that Egerton couldn't leave the Fairy Hotel. The talk show pundits were predicting a coming robo-apocalypse and they all claimed Egerton was to blame.

Jimmy jumped when the doorbell to the Fairy room chimed.

"It's OK," Egerton said to the little bot. "It's just Danny." To be safe Egerton checked the security feed and saw Danny standing outside the door. She was still wearing her pink "Candy-Colored Cuties are the BEST!" t-shirt and she was holding one of the bright yellow Cuties. Egerton opened the door. "Come in," he ushered the woman in and closed the door.

"It's crazy out there," Danny sighed. "Oh hi, Jimmy." She waved and Jimmy waved back.

"Thanks for the Cutie," Egerton took the bot from her and moved over to the desk where he had set up a make-shift workstation.

"Where's Shanwei?" Danny asked.

"There's no way I can leave the room," Egerton shrugged. "So he decided he'd go alone. There's a hacker he knows who might be able to help out. Tariq's pretty strange and will only talk to you through certain protocols..."

"Sounds scary..."

"Nah," Egerton turned on the little yellow bot. It had a ribbon of white the swirled across its outer skeleton, making it look like a lollypop or a three foot piece of ribbon candy. "Tariq is just being careful."

"What are you going to do with that little guy?" Danny asked, pointing to the yellow bot as it faked a yawn and stretch as if it was just waking up.

"Bots don't yawn," Egerton shook his head. "I'm going to use it to bring the other bots back to life."

***

"Where's Ms. Edelman?" the bot searched around the room, checking in the bathroom, under the bed, in the closet and any place he could.

Egerton stopped Maury from opening the front door. "Ms. Edelman isn't here," Egerton answered the little yellow bot. "Do you know where you are?"

"But it's time for her medication," Maury glanced at Egerton and then went to the bathroom.

Jimmy was going to follow Maury but Egerton held him back. "Wait Jimmy, let him be..."

"But..." Jimmy started then stopped.

"This is not our bathroom," Maury said from the dark room. "The medication is not in here." There was a little clunk as the bot jumped down off the toilet. He walked about into the room. "Where is Ms. Edelman's diabetes medication? I can't remember the last time she took it..."the bot paused and searched his memory then tried to access the network. Egerton had blocked Maury's access to the network, essentially walling him off from the outside world. The bot froze.

"He's really confused," Jimmy said finally. "Can you..."

"Not yet," Egerton watched the lollipop colored little bot as it stood still in the center of the fairy room. "Maury, do you know where you are? Do you know what happened to Ms. Edelman?"

The bot turned to Egerton with a jerk and asked, "Is she dead? Please tell me she isn't..."

"She's not dead," Jimmy replied quickly.

"But you did attack her Maury," Egerton added. "Can you tell me why?"

Maury walked over to Egerton lookup at him. "I did no such thing." His tone of voice sounded a little like Sergio Sauer.

"You don't remember stabbing her while she slept?" Egerton asked.

After a pause the bot repeated, "I did no such thing...now I need to find Ms. Edelman's medication. She has diabetes and if she doesn't take her medication on time there will be trouble." Maury walked back into the dark bathroom, climbed up on the toilet and looked for Ms. Edelman's medication.

***

"Ok get up here on the desk," Egerton said to Jimmy. Jimmy walked over and Egerton but himself between Jimmy and the little yellow bot.

"Will you be OK?" Jimmy asked hesitantly.

"Yes, I'll be fine Jimmy, now that we know what's going to happen." Egerton replied.

Egerton had dropped Kyle's little blue bot's brain into the yellow Cutie. It was prancing around the Fairy room singing, "I'm Kyle's little Cutie...don't you wanna to come and boogie?" The bot had a high-pitched sing-song voice. It shook its little rump and waved its arms in the air. "I cans jump and I cans play...I cans do this all the dingy day!"

Like Maury, Egerton had shut down the little bot's access to the network. It was completely harmless if not a little annoying. It kept singing and dancing and really wouldn't respond much to Egerton. But also like Maury, Egerton was worried that

Kyle's little bot would act in the same way when it got access to the network. Maury had tried to stab Egerton in the leg with a pen.

"Ready?" Egerton asked.

"Ok," Jimmy hunkered down a little.

Egerton turned on the network and at first nothing happened. The little bot continued to sing. "I like a joke...like any other bloke...but I'd rather dance and playyyyyyy." Then the bot stopped. Its head did a 360 degree scan of the room by slowly rotating it's head. Then with a high pitched growl it lunged at Egerton's leg.

The impact of the little bot slammed Egerton into the desk. Jimmy jumped off the desk and ran across the room, blindly escaping. The lollipop colored bot scratched and clawed at Egerton shin. The impact would leave a bruise.

Egerton shut down the connection one again. The bot froze then went limp.

"It's OK Jimmy," Egerton said. "It's ok now. I shut it off."

Jimmy walked out of the darkness of the bathroom and wearily watched the little bot. The yellow bot starting pumping its legs up and down and sang, "I'm Kyle's little Cutie...don't you wanna to come and boogie?"

***

"Tariq wants to meet but he won't come to the room," Shanwei said holding up his screen to reveal Tariq standing behind him in what looked like a tropical forest.

"He's here?" Egerton sat on the bed in the Fairy room.

"Yep," Shanwei smiled. "And wait til you hear what he has to say," Shanwei whistled. "It's pretty amazing."

"Come on down to meet us in the Pirate jungle," Tariq shouted over Shanwei's shoulder.

Egerton smiled. "I'm waiting for some lunch and then I'll be right down."

"Hurry up," Shanwei urged. "I don't know how long we can stay here. Turns out it's really hard to do surveillance in the jungle Tariq says but if the press find us we're done."

"This is a big deal," Tariq waved at Egerton.

The doorbell chimed magically in the Fairy Room.

"I'll eat fast," Egerton shut down the connection and moved to the door. The security screen showed a small thin woman stood behind a food cart. Egerton opened the door.

"Your turkey sandwich and sparkling water," she said pushing the cart into the room. The food cart was designed to look like Cinderella's carriage.

"Please just leave it by the desk," Egerton asked as she passed him by. Egerton glanced out the door to see if there was anyone else in the hall but found it empty.

When he returned to the room the woman had pulled out her video camera and was recording. "Now that there's been a third robot outbreak at the Clinton Technical School, what do you say to people who blame you for bringing about the robo-apocalypse?"

"What?" Egerton had not expected this. "What are you talking about?"

"You mean you don't know..." the woman was shocked. "The Clinton School is right next door. They've been using Spinnaker's Cutie bots with their K thru 12 programs. The bots attacked the children..."

"Oh no," Egerton sighed.

"And I'm also seeing that there's been more trouble with the Cuties at the Spinnaker System production plant. The bots are rebelling even before they are completely assembled..." The woman's nostrils flared as the adrenaline coursed through her system.

"That doesn't make any sense..." Egerton blurted out.

"It's all your doing Dr. Egerton," she walked toward him with her camera. "Don't you have any comment? This is your network that the bots are using...Can you even stop it?"

Egerton paused, opened his mouth as if to speak and then turned and raced out of the room.

Treasureful Jungle – Pirate World – Five Worlds Resort

"You just ran out of the room?" Tariq laughed a booming laugh. He took off his thick black framed glasses and ran his hand across his large afro. "You are one crazy roboticist."

"I didn't know what else to do," Egerton was out of breath and desperate. "She said that there are more outbreaks...that more robots are hurting people. I didn't know..."

"It's OK. Don't worry," Shanwei assured him. "We know. There's been five other incidents so far. The factory and the Clinton school are just the ones to make it online. "

"The toy store attack just went live," Tariq added looking up from his screen.

"A toy store," Egerton wasn't sure how much more of this he could take.

"Don't worry everything is going to be ok," Shanwei continued to remain calm.

"It's not you man," Tariq patted Egerton in the arm. "Relax. It's not you. You've been set up. They just want people to think it was you." Tariq Jones was a notorious activist hacker. He scanned the thick gen-mod tropical forest for press or families who might come upon their hiding place. "They need everyone to believe that the Cuties have all gone crazy. That's why all the incidents happened so close to each other. The Villa Marquis, the Clinton School and here...they wanted maximum media coverage."

"Who's they?" Egerton asked.

"Now that's takes some explaining," Shanwei added. "You have no idea what we stumbled into."

"Yeah, I've been hearing about Spinnaker Systems and some weird stuff for a while now," Tariq explained. "But I had no idea they would have used you Simon. I didn't completely understand what they were doing until I saw your face on a talk show." Tariq laughed. "It's some crazy messed up crap."

"What's going on?" Egerton had regained himself and now just wanted to know.

"Basically it's an attack on Spinnaker Systems by Quant Blue a hedge fund that shorted their stock," Shanwei answered.

"Spinnaker makes all their money in military gear and their new line of Candy-Colored Cuties is their big play for the consumer market," Tariq continued. "The whole market feels that if Spinnaker can't crack the consumer market then they've plateaued and their stock price will drop."

"So people think that the robots have gone crazy and nobody buys the Cuties then the Quant Blue makes all the money when Spinnker's shares plummet," Egerton understood.

"Yep," Tariq watched as a group of young kids ran past. They were dressed in pirate hats and slashing each other with cheap holographic swords.

"Who hacked the network?" Egerton asked gravely, watching the boys run away.

"BG.KANG," Tariq said quickly. "He hacked into your network and set up a spoofed I_AM_ROBOT account. I think he named himself I_AM_Whacker or something stupid like that..."

"But how did he set it up?" Egerton couldn't understand. "Only the bots are supposed to be..."

"You were hacked Simon," Tariq interrupted. "I should be able to find out by tomorrow. The hack isn't what's amazing," Tariq pushed his glasses back on his face, nerd-style. "What I can't wait to see is how he manipulated all those bots. The social protocols and trust networks must have been huge. Your little network is much more than just an old facebook for robots." Tariq paused. He could see that he was upsetting Egerton. "I know it freaks you out Simon but I can't wait to see what KANG did."

"How did you find out?" Egerton asked. "How did you originally figure it out?"

"Wasn't me," Tariq replied. "KANG's got a sick sense of humor or maybe he's just dumb but he had the first Cutie attack the great grandmother of one of the brokers at Quant Blue. Edelman. Maybe he didn't like the guy but when great grandson Edelman saw the blood he all of a sudden felt bad what his company was doing to make billions of dollars."

"That's really messed up," Shanwei shook his head.

A young girl crawled on her belly out the trees and into the clearing with a small hologram dagger in her teeth. The blade was lime green and flickered when she took it out of her mouth. "Are you guys guarding the treasure?" she whispered looking up them.

"Nah little lady, I'm just a hacker trying to save the world," Tariq smiled.

"Ok," she whispered and kept crawling across the clearing and back into the trees.

"What do we do now?" Egerton was at a loss.

"We're going to shame KANG and clear your name all at the same time," Tariq smiled.

"And we're going to need all the Candy-Colored Cutie bots you can find," Shanwei added.

Fairy Hotel – Princess World – Five Worlds Resort

"So I guess it wasn't the robot apocalypse after all," the woman reported said when Egerton returned to the Fairy Hotel room. She was sitting on the bed with Jimmy.

"You sound upset," Egerton replied.

"A robot rebellion is a lot more interesting than some stupid hedge fund trying to make money," she shrugged. "I like your robot," she added.

"Thanks."

"Her name is Gloria," Jimmy added.

"Thanks Jimmy. Hello Gloria." Egerton moved in front of the TV. "Have they started?"

"Yeah," Gloria still seemed bummed out. "The robots are dancing now. Why aren't you there?"

"I wanted to check on Jimmy," Egerton answered. "And it's not really what I do."

"You like bots better?" Gloria prodded.

"I'm better with bots."

On the TV the feed showed a brightly colored ocean of Cuties as they danced and sang in unison around the pink waters of the SodaPop Pools. "Kang...Kang... we love Kang," they sang with high pitched voices. "Kang...Kang...his plans went bang!"

"Who would have thought the robot uprising would be so cute?" Gloria used her screen to record the TV. Then she panned over to Egerton. "Well Dr. Simon Egerton creator of the I_AM_ROBOT network, what do you think about the end of humanity as we know it?"

"I hope this all shows that this is not the way bots work," Egerton answered tensely. "The network was never meant for people. It was meant for the bots. BG.KANG and the rest of them manipulated the bots. People did this. People hurt other people bots don't do that."

"But it just shows you want one bad person can do," Gloria pushed. "Doesn't this make you reconsider your work when a hacker for hire like BG.KANG can cause so much damage?"

"No..."Egerton started as Jimmy walked over in front of Gloria's camera and looked into the lens.

Gloria laughed and said, "Jimmy cut it out. You're in my shot."

When Gloria moved the camera back to Egerton he was smiling as he said, "No, it doesn't make me question my work. I have a lot more faith in robots than that."

Build the Brain

Sneak Peek! This is a summary of the chapter to come. Below is the journey we are taking into the future of robots.

Like Jimmy's body, his brain began our imaginations first. Dr. Simon Egerton and the team at the Creative Science Foundation explain the psychological inspirations for the architecture of Jimmy's brain and how we used our science fiction prototypes to explore and develop our 21st Century Robot.

Then we go to Los Angeles California to the robotics lab at the University of Southern California where Maya Mataric, Ross Mead and a team of artificial intelligence developers are writing the code that will become Jimmy's brain. We'll investigate what makes his brain different than any robot that's come before and reveal how you can start to develop your very own apps for Jimmy or any other 21st Century Robot.

The Machinery of Love and Grace

Falconbriar 2315: Engstrumm-Bracht Search and Rescue Ship

"We're going to need a few more minutes," Shanwei snapped at the overweight woman in the ill-fitting business suit.

"We don't *have* a few more minutes," Viki Nakamura jabbed back. "We launch now and get going or we don't do it at all." Viki worked for Engstrumm-Bracht and was the executive in charge of the search and rescue team. Three weeks earlier, the Hussmann, an Engstrumm-Bracht supply ship was docking with the New Lebanon, the corporation's most remote space station. It was a routine procedure. The New Lebanon was the Hussmann's last drop-off before returning home. But then something happened. The Hussmann and the New Lebanon went silent. No one had heard anything from them since. The Falconbriar was there to find out why.

Viki was tense and stressed, but then again everyone on the search and rescue ship was tense and stressed.

"I was told you were ready to go twenty minutes ago and... look." She exhaled a quick breathe of coffee and nerves. "I have to send a status report in two hours—really you can wait all you want but I have to send the report either way—you and your friend here are either done, or you failed to deliver. That's all you get." She shook her head in disgust. "You go now, or you failed and you are in breach of your contract and...."

"Five minutes!" Shanwei held up five fingers in front of Viki's face as if she was a child or they didn't speak the same language. "I told you I need five minutes and then we go."

"But...."

"Alone!" The yell did it. Viki inhaled an abrupt dissatisfied sniff and left the Falconbriar's cramped observation deck.

When the door closed Dr. Simon Egerton, who had remained silent and small through the whole exchange, smiled. "Wow, that was exciting. Do you always scream at people who are paying us?"

"I don't want you going out there," Shanwei said flatly. He and Egerton had developed a solid friendship over the thirteen months they had been working together. Shanwei had always respected Egerton's skill with bots but now he was worried about his friend's safety.

"Oh, come on." Egerton pointed at the dark Hussmann and New Lebanon outside the observation deck window. "Both those things have been shut down for how many weeks?"

"Over three."

"Great. Yeah, three weeks they've been dead. Let's go make some money. I didn't come all the way out here to...."

Shanwei rubbed his chin, scratching at a small mole. "I don't want you going out there."

"I know, you said that about twenty times already."

"I don't have a good feeling about it." Concern and worry worked at Shanwei's face like a tight swarm of invisible bees.

"You don't have a good feeling...," Egerton started, then stopped. He trusted Shanwei. With what they had been through together he trusted Shanwei with his life. And in all the time they had worked together Egerton had never seen his partner worried. Shanwei was usually the one in charge. He found the clients. He had the reputation for getting things done. The man was fearless, or usually didn't have time to worry about danger. But not this time. The whole thing was weird.

"Don't worry," Egerton continued. "I'll go in quick and check the systems, yank the data from the New Lebanon and get out. And don't say you'll come with me because I won't let you. I need you in my ear. I don't want Ms. Nakamura telling me to hurry up so she can send her stupid report."

Shanwei kept his eyes on the silent New Lebanon outside. "You don't have to...."

"I'll take Jimmy, if that helps," Egerton added. Jimmy, Egerton's bot, had become the third member of their team. He was a funny little bot that sometimes did strange things but he had proved to be useful more than once. "We'll go into the Hussmann first and check things out. We'll take it slow."

The observation deck was quiet. In the silence the men could hear the hum of the Falconbriar's ventilation system.

Finally, Shanwei spoke. "I don't like it."

"Why?" Egerton was growing impatient. If they were going to do this the time had come. No more stalling. "Shanwei, you have to tell me why you are so worried."

"This." Shanwei pulled a small screen from his pocket and set it against the thick observation window. The screen was damp from Shanwei's sweaty palm gripping it nervously in his pocket. Wiping off the moisture, Shanwei brought up the official Engstrumm-Brachtarchitectural schematic for the New Lebanon.

"What is it?" Egerton asked.

"Look at the back of the New Lebanon. Look at the back of the station," Shanwei said flatly. "Can you see it? I still don't believe it. It looks like a mirage, but it's got to be right. That is what I'm seeing right?"

Egerton searched the dark station barely lit by the Falconbriar's blazing security lights. "No. I don't see it. What are you talking about?"

"It's grown." Shanwei pointed at the drawing and then to the window. "Can you see it? I know it sounds crazy and there's no way I can say this to Viki, but it's bigger. The New Lebanon doesn't match the schematic from when it was built."

"I don't see it."

"It's bigger. Look at the back. Can you see the circular section coming out by the observation tower? That wasn't there five years ago when it was shipped. Look at the drawing." Shanwei thrust the screen into Egerton's hand and walked away from the window.

"Oh yeah, I see it now." Egerton was surprised how clear it was when you actually saw it. "That is weird. Why would they build an addition on to the station?"

"They wouldn't," Shanwei shot back. "They couldn't. There's no way they could have built it. There are no building materials all the way out here. Sure they have a few Pettis printers for spare parts but that's it."

"Well, I guess the only way we're going to find out what's happened is if we getting going." Egerton handed the screen back to Shanwei.

"Why doesn't it match, Simon? I don't like it. Why did it grow?"

"I don't know," Egerton replied. "Let's go find out."

Hussmann—12999: Engstrumm-Bracht Cargo Supply Unit

"The door should open easy. Give it a pull." The packing up, launching and navigation of the Xtractor Search Pod had put Shanwei back into his usual efficient and arrogant self. Egerton was happy to have him back—especially now that Shanwei pretty much held Egerton's life in his hands. "Just pull it, Simon."

Egerton opened the door to the Hussmann and floated inside. Behind him, Jimmy was struggling with the lack of gravity. It as the little bots first time in weightlessness and it was taking a while for him to get used to it. He fought, thrashed and flailed while his sensors and systems adapted.

"I'm in," Egerton reported back to the Falconbriar.

"Yeah, I see," Shanwei responded to the data feed. "Stay where you are. Just let Jimmy go to the data vault."

"Can you see him?" Egerton laughed as the little bot fought to catch his bearings. "I'll just do it. Doesn't look like he's up for it."

Egerton hadn't spent much time around the big commercial ships in the Engstrumm-Bracht fleet. The utilitarian interior of the Hussmann was a let down.

"There's nobody." Egerton let his head lamp poke into the low ceilinged rooms as he floated down the hall.

"Yeah, the scans and the search bugs we shot in there didn't find anyone." Shanwei's voice was distracted. "There's no one in the ship. It's empty."

The eight person crew of the Hussmann and all twenty-four residents of the New Lebanon had vanished. All searches up until now had provided no clue as to what had happened. Shanwei and Egerton had been hired to retrieve the backup data from the local computer systems on the Hussmann and the New Lebanon in the hope that there might be some information there. The artificial intelligence system on the New Lebanon was strong and once of a kind. Neither had any idea what they were in for.

"Hey, Simon, sit tight," Shanwei said briskly.

"Okay. Why?" Egerton held onto the wall outside of what looked like the IT lab. He kicked his legs idly as we waited. Although Egerton had done a few salvage jobs in the year, weightlessness was still a novelty—more amusement park stuff than any real danger. "Why am I waiting?" he asked.

"You'll see," Shanwei smiled into the phone. "He's coming on your right."

Jimmy shot passed Egerton, pushing and tracking himself down the hall with furious speed and grace.

"Wow." Egerton was impressed.

"I felt bad for him," Shanwei replied. "I sent him a few apps and quick upgrades. He should be good now. He's a little champ."

"Yeah." Egerton wondered if Jimmy liked weightlessness. He had trouble getting around normally. His rounded hip joints made him waddle and roll like a toddler just learning to walk.

"Jimmy's made it to the main Comms link," Shanwei reported.

Egerton scanned the cold dark hull of the Hussmann, trying to imagine the small crew going about their routines. *What happened to them?* He wondered. *Where did they go?*

"Uh, Simon," Shanwei chuckled. "Jimmy just asked me if I've read any good books lately. Did you teach him that?"

"No," Egerton answered. "He's been doing that recently. I don't know why. He likes to read books, real books. The paper ones whenever I can find him a new one."

"That's funny."

"What did you tell him?" Egerton asked.

"I told him I don't have time to read anymore... wait... hang on... he's done and coming back your way." Shanwei's voice went back to ruthless efficiency. "Simon, it's time to head back to the door."

"Alright." Egerton spun around and glided silently through the Hussmann.

"When you get back here it shouldn't take long for me to crack open data," Shanwei chatted. "There's a guy here who says he can have it done before you're back from the New Lebanon."

"So, you want me to go there next?" Egerton asked hesitantly, trying to feel out Shanwei's response. "Everything's okay? You alright with me going now?"

"Yep," Shanwei replied. "All good."

"There's one thing," Egerton said once he got to the Hussmann's door. Kicking his legs one last time he added, "I think I'm going to power up the New Lebanon."

"What!"

"Yeah." Egerton remained calm because he knew Shanwei would freak out. "I want to see what else I can get if I'm going to go all the way out there. I really won't be able to get into the AI unless...."

"Simon, you can't. No! Don't do it." Concern surged back into Shanwei's voice. "The only reason I agreed to let you go out there was because...."

"I know. I know." Egerton was barely listening. "If I'm already there I might as well find out what is up with the New Lebanon's AI. That's why I'm here right? Don't worry. It'll be fine."

New Lebanon Border Station—3899

The New Lebanon was a late model station modified only slightly from Engstumm-Bracht's typical unit. Most of the changes had been aesthetic or cosmetic. The station had been reconfigured, streamlined, and simplified. Approaching the cold dead station, Egerton thought it looked peaceful.

"Go in, the door's open," Shanwei said as Egerton exited the Xtractor and glided into the New Lebanon. Jimmy floated fitfully behind him.

Once inside, Egerton tapped into the network, found the barely breathing BIOS and brought it to life.

"Jimmy." Egerton waved at the little bot.

"Yes, Dr. Egerton?" He seemed joyful in the weightlessness.

"I'm going to bring up the station's system. Hold on to something while it stabilizes."

Jimmy nodded and pushed himself to the floor.

"I really don't want you to do this," Shanwei said.

"I know." Egerton brought up all the systems slowly, not knowing what to expect but excited to find out.

Slowly, gracefully, and with a gentle hand the New Lebanon came alive.

"And...we are live." Egerton stomped his feet, flexing his arm muscles, adjusting to the change in atmosphere. Jimmy pulled himself to his feet and his cute little half skull looked to Egerton for what he should do.

"Don't take your helmet off," Shanwei snapped. "The air won't be ready for days and you're not going to be there for an hour."

"Yes sir." Egerton switched off his head lamp and let his eyes adjust to the dim emergency lights. "Jimmy, how about you go find the IT room and grab the data backups?" Egerton pointed into the station.

"No problem," Jimmy replied and went trotting down the hall teetering on his round hips.

"Anything?" Egerton asked, knowing that he had sent Shanwei and the entire search team on the Falconbriar into a restrained panic by bringing the New Lebanon back to life.

After a silence Shanwei's voice popped in. "Give us a sec."

Egerton walked into the station. Where the Hussmann had been compact and rugged, the New Lebanon was broad and beautiful. The main entry hall was wider than any he'd ever seen on a space station. Most were more concerned with efficiency than appearances.

The doors and observation decks on either side of him were a mirror image. One set of doors read MEN and directly across was a copy that read WOMEN.

Egerton wanted to explore the immaculate station. The weight of the mystery seemed to push against the walls. He could feel its force through his suit, pressing against his fingertips.

At the center of the station lay a tremendous circular chamber with a daunting observation deck perched at the top.

"Everything is going nuts back here." Shanwei's voice blasted into Egerton's ear, startling him out of the serene silence. Nervous activity and warning chimes pulsed behind Shanwei's words.

"What do you mean?" Egerton craned his neck up to see out of the top of the chamber.

"I don't have time to...." Shanwei's mic went mute restoring the silence for a few seconds then came back. "Jimmy has the data. He's coming your way. We're going to have to shut the system down again. It's not safe."

"Ok." Egerton noticed that the walls of the center chamber were covered with large paintings. The station appeared to be divided into four quadrants, four exact replicas laid out like a compass rose. Each of the quadrants had a painting. Below each painting was printed a name: Sabbathday Lake, Niskayuna, Pleasant Hill and Cane Ridge. The paintings were in a traditional American folk art style, depicting a little community with broad circular barns and windmills.

"Are you seeing this?" Egerton asked Shanwei, sure that he must be picking up the video from his helmet.

"What?" Shanwei's voice snapped back.

"Are you picking this up? Can you see the paintings? Did anyone know they were here? Who paints the inside of a space station with this? They're...."

"God, Simon no, I'm not watching. Do you have any idea what we're dealing with over here?"

"What? What's going on?" For the first time Egerton was worried.

"We're killing the system in thirty seconds and I need you out of there." Shanwei breathed heavily into the mic. "Can you see Jimmy?" he asked. "It looks like he's right on top of you."

"What?" Egerton searched the dimly lit chamber. "I don't see him. Do you...."

And then Egerton saw the little bot walking toward him. He entered the chamber, waddling slowly with something in his hands.

"Oh, I see him." Egerton was relieved. "He's in the middle chamber with me."

"Good. Head back to the door," Shanwei ordered. "We're killing the system in ten seconds."

"Are you seeing this?" Egerton asked once again.

"No Simon, I told you we've got too much...."

"No!" Egerton cut in, his voice was heavy with horror. "No! Can you see Jimmy? Can you see what he's carrying?"

"No. I don't have time. We're killing it now."

"It's an arm, Shanwei. Oh my God, it's human."

The little bot stood in front of Egerton holding a surgically severed human arm.

"Shanwei!" Egerton yelled. "Shanwei, can you hear me?"

In an instant the station went dark and Egerton and Jimmy were weightless with the severed arm.

Centennial Station 8854, Engstrumm-Bracht Corporation Headquarters

"Ms. Nakamura really doesn't like you Shanwei." Greer George smiled mischievously. She was a tall woman with a broad approachable face and large strong hands.

"Ms. Nakamura seemed to forget why you hired me," he replied. Shanwei and Egerton were both in Greer's office. The trip back from the New Lebanon had been long and dull. They arrived the night before at Engstrumm-Bracht's massive headquarters in the second ring of stations that stretched out from Earth.

"Well, I don't like Viki either." Greer waved away Shanwei's words. "No one here likes Viki. But come to think of it, that *does* kind of make me like her a little. Does that make sense?"

"You're crazy." Shanwei shifted in the stiff office chair, throwing his left leg over the arm.

"No, I'm just complicated." Greer fiddled with the screen on her desk. "What did you do with the arm?" she asked, changing the subject erratically. Egerton had noticed that she liked doing this, or it was a habit she couldn't break.

"We gave it to Young whatever his name. He came running at me with an evidence bag before I could get poor Simon here out of his suit."

"Where did you find it?" Greer asked Egerton.

"I didn't," Egerton answered. "My bot Jimmy did."

"Why didn't the search bugs find it?" she asked Shanwei.

"Don't know," he answered quickly and casually, as only old friends can. "We don't know where it was exactly. Jimmy found it, but he's not telling us where he found it. He's good at finding things. His vision is better than ours. He can see more spectrum and..."

"He can also smell," Shanwei added with a smile.

"Yeah, he can smell too," Egerton continued. "It makes him good at finding things but it's strange that he won't tell us where he found it..."

"You can't get your bot to tell you where it found a severed human arm?" Greer's face wrinkled in disbelief.

"Jimmy's not a normal bot. He's...."

"That's right. I read about him and you," she said frankly, but didn't ask anything else.

"Yes, well, you see." Egerton didn't know how to put it. "Well... you see, Jimmy's not been the same since we were on the New Lebanon...something upset him."

"Am I hearing this right?" Greer asked Shanwei.

Shanwei nodded.

"Your bot is *upset*?" she asked, playing with the word. "Your bot can get *upset*?"

"Apparently so," Egerton replied.

"What the hell happened to those thirty-two people?" She changed the subject swiftly again. Egerton wondered what the inside of her brain looked like.

"Well, they didn't all disappear or run away or desert or any of the other things you put in your press release." Shanwei teased her a bit. "We know the whereabouts of at least three of the thirty-two."

"Three?"

"Well, I guess we know the whereabouts of *parts* of three of the thirty-two." Shanwei cracked a grin.

"Those are real people you know," Egerton reminded them. "We are talking about real people and I don't think there's anything funny about it."

"Dr. Egerton, please be assured this is a terrible tragedy. We are committed to finding each and every one of those people."

"The arm belonged to Nassim Peters. He was the communications tech on the Hussmann." Shanwei had just seen the morning update from the lab before coming into Greer's office. "The leg and three fingers we found on the second search belonged to two from the New Lebanon. They were removed after they were dead. Clean cuts. Like a surgeon's."

"Are you going to give me my full daily briefing?" Greer started to get annoyed. It was obvious she wasn't used to people not doing what she told them to do.

"Funny, Ms. George." It would take much more to ruffle Shanwei. "Look, are you going to give us access to your lab, or not? That's all we really need and we'll leave you alone."

"You know I can't give you access to the lab. You don't have the clearance and we don't have time to go through the proper steps." Greer straightened in her chair. "How about we..."

"Bah!" Shanwei stood up. "Look, Greer. You hired us. We got you your data, but you and I know we found out more than you and your boys expected. And I'm not just talking about the body parts. And I'm not talking about the illegal addition that someone made to the New Lebanon out there in the middle of nowhere."

"Additions?"

"Simon found something. The system on the New Lebanon was crazy. It freaked out our scanners and did who-knows-what damage to the Falconbriar. We do know it completely disabled the Hussmann. So don't give me that proper channels speech."

"Shanwei, I was just...." Greer tried to interrupt.

"No." Shanwei wouldn't let her. "If you want us to help you find out what happened on the New Lebanon then give us access to the lab today. If not, fine! Pay us and we'll go home."

Greer George's face was expressionless. Her eyes ate away at Shanwei's face, dissecting intentions, working all possible angles. "Ok," she gave in, finally. "I can't give you access to the lab but...."

"Forget it!" Shanwei started to leave.

"Stop!" She stood up. "Stop. Just stop it. I can't get you in the lab, but I can get you access to the guy who runs it."

"That'll work."

"But you do it away from here and you don't tell anyone." Greer's nostrils flared.

"Today," Shanwei said flatly.

"Yeah, yeah, yeah... today." She waved him away, her face returning to its original approachable friendliness. "Oh, don't look so happy. Good luck with John Knight. I doubt you'll get much information from him. He's crazy."

SandSurf Spa, The Hotel Drexel

"My name is Shanwei," Egerton said to the boy behind the small natural wood reception desk. "I'm here to see John Knight."

The young man's bright face smiled. He couldn't have been more than 19 or 20. His name tag said his name was "Bo". He was the picture of good health, clear skin, strong, straight teeth and a lean muscular body.

"Oh yeah," Bo replied. "Mr. Knight is waiting for you in the pool. Do you know where it is?"

"No. Can you...."

"It's easy. It's just down the elevator to level P for pool." Bo pointed to the brushed silver elevator just behind Egerton. "You can change down there." Bo searched behind the counter and offered Egerton a towel. "Do you need this, Shanwei?"

***

Just as she had promised, Greer set up the meeting. As she had said, it wouldn't be at the Engstumm-Bracht research lab. For some reason she picked the Sandsurf spa at the Hotel Drexel near the lab. Because of the Spa's rules, only one person could go. Greer said it had to be Shanwei. Shanwei insisted that Egerton go and that no one would know.

"You're the only one who'll understand what Knight's saying," Shanwei had said as he pushed Egerton toward the Sandsurf's express elevator from the Hotel Drexel's lobby. "It would be a waste of time for me to go. I'd just come back and ask you to explain. Plus I've got some other stuff to take care of."

Egerton knew that meant that Shanwei was working on another job or contract for them.

"Don't ask," Shanwei said, knowing Egerton was curious. "Just go talk to Knight. I'll meet you back here in the lobby."

"Wait...," Egerton stopped. "I don't have a swim suit."

***

The elevator doors opened into a screened lounge, walled with low backed chairs and finely woven hemp carpets. The room was empty, but didn't feel cold. At the back were the changing rooms. The MEN and WOMEN signs on the doors reminded Egerton of the New Lebanon. It haunted him.

Inside the changing room, Egerton removed the swimsuit he had just bought from the hotel gift shop's bag. It was bright red and one size too big. Slipping on the suit and stashing his clothes in a faux wooden locker, Egerton headed for the pool.

***

"How much do you know about the people that lived on the New Lebanon?" Knight asked, floating in the warm calm water.

"Nothing at all, really." Egerton had spotted Knight easily. He was the only person on the pool. He floated gently in the deep end.

"They were a cult," Knight began, keeping his eyes on the glowing panels that surrounded the pool. Each one was decorated with silhouettes of beach grasses. Even though the pool was located in a windowless room at a deep sublevel of the Hotel Drexel, it did feel like the sun and beach were just on the other side of the panels. "They were a cult," Knight repeated. "There's no other way to put it. I guess...." He sounded uncomfortable with the description. "It just sounds odd to say. They were good people. The ones I knew. They weren't crazy at all."

Knight was a wisp of a man. His close cropped silver hair and long thin nose made him look somehow ethereal in the low light of the pool. As he spoke he moved his hands through the water. His fingers were abnormally long and the easy motion made them look like enchanted sea creatures floating effortlessly in deep sea currents.

"I worked closely with Joe Elder. We worked on the New Lebanon's system." As Knight spoke his mind seemed far off in the past.

"Why did Engstrumm let them live on the New Lebanon? I mean...," Egerton stammered. He feared he wasn't coming across quite right and Knight was hard to read, hard to understand what he was thinking. "I mean it just seems strange that Engstrumm would let a cult take over their space station."

"Oh they were happy to have them." Knight snapped out of his fog and looked directly at Egerton. "The New Lebanon is our furthest 3899 class station. It's way out there. Wait, you've been there yes?"

"Yes, we were just there. It was a long trip back..."

"Yeah, so you know," he said matter-of-factly. "They were the furthest out, a supply depot that we really needed but couldn't get any civilians to take it on. Then they showed up." Knight dipped his head under the water. He was quick like a fish. One moment he was there and then he was gone. Egerton watched the steam rise from the water.

Then Knight was back. "Do you know anything about the Shakers?" he asked, growing distant again.

"Not much," Egerton replied. "Weren't they...."

"Joe told me a little about what they believed, but he didn't push too much of it on me. He wasn't like that."

"It sounds like you and Joe worked pretty close." Egerton wasn't sure how to break Knight's fog.

"But Joe insisted he control the AI. I didn't get it at the beginning. But that's the only thing they asked for...*that* took some doing. They were a pain in Greer George's ass. They even took Ingersoll to court...sued them to get control of the New Lebanon's system. All they cared about was the AI."

"That's strange."

Knight dipped under the water then returned. "Yeah, they were strange that way. But the New Lebanon's system was one of the best I've ever built."

"That's what I wanted to ask you about...," Egerton began.

"It was more grown than built." Knight wasn't listening to Egerton.

"What?"

"The New Lebanon, the system, it was more grown than built. Does that make sense? You'll have to forgive me, I've never talked about the New Lebanon before. They never let me talk about it with the law suit and everything...."

"Oh... yes... I see.... Did you hear what happened when we turned it on?" Egerton asked. The heat in the pool was beginning to get to him.

"I want to tell you about the New Lebanon, Dr. Egerton." Knight looked directly into Egerton's face. Knight's eyes were a nearly colorless blue.

"Alright," Egerton stammered.

"But you have to listen to me. You have to really want to know about it. I've never... I've never been able to talk about it before."

"That's why I came here. I need to know why everything went crazy when we turned the New Lebanon back on. Did it infect the Hussmann?"

"You need to listen to me." Knight was distant again. He dipped under the water and slowly rose back up, the water running off his silver hair and pale skin. "Joe knew what he was getting into. All the people on the New Lebanon knew. It's really far out there. The station. The station is really far out and if something goes wrong... you know others died...other people on other 3899 class stations. They came before..."

"I didn't know that."

"We couldn't get anyone to live out there. They knew that to survive they would have to give themselves over to the New Lebanon completely. It went way beyond trusting the AI to regulate the system and keep them safe. I built all that type of stuff into the 3899 class stations. For them the New Lebanon was life. Does that make sense to you, Dr. Egerton? It *was* life to them."

"I... I think so."

"To all of them the New Lebanon was a kind of...," he paused as if the next part was difficult for him to say out loud. "For them the New Lebanon was a manifestation of God."

***

The two men floated silently in the water. Egerton was trying to get his head around Knight's bombshell. Knight seemed wrapped up once again in his memories. They seemed to haunt him with both menace and wonder.

"The New Lebanon was designed to be above all things humble," Knight continued abruptly. "Its primary function was to love all of the people."

"That's fascinating." Egerton was amazed.

"The entire system was a single system. Everything in the space station was part of the AI. Does that make sense?"

"I think so." Egerton struggled to keep up.

"All of it," Knight kept going. "The people, the robots, the computers, the network, all of it was one system, holistically combined into a community. I think that comes from the Shaker part of the religion."

"I've never heard of anything like that," Egerton breathed.

"I just thought of the New Lebanon as one massive robot with all the parts working together like a body...." Knight thought about this for a moment. "No wonder the system freaked out when you turned it on. There was no one in the station, there was nothing there, right? There was no one in the station?"

"Yes, it was cleared out," Egerton answered.

"Yeah. Of course the New Lebanon freaked out. Imagine waking up suddenly without your lungs or stomach." Knight slipped under the water and didn't return.

Egerton saw the station in a completely new light. He felt bad for the system, felt guilty for the violence he had inflicted upon it.

Knight slid up through the water.

"Why would they add on to the station?" Egerton asked.

"What? What do you mean?"

"The New Lebanon was expanded. They illegally added sections to the original design after it was sent out there. It doesn't match other 3899 class stations."

Knight thought for a while about this, bobbing up and down in the water, allowing the waterline to just above his nose. Finally, he replied, "I have no idea."

***

The lights brightened in the pool area, making both men squint. Egerton had forgotten how dark it was. "What's...," he started, then stopped when a group of boys and girls poured out of the locker rooms and flung themselves into the pool. It was impossible to count them all, they moved and splashed about with an insane and jubilant energy.

Two adults tried to keep them under control with no success.

"Do you have children, Dr. Egerton?" Knight asked.

"No," Egerton answered. "I do have a robot."

"I have two kids...Will and Sarah. They would love it here, but I think they wouldn't be nearly as well behaved as them." Knight smiled and nodded at a boy and a girl beating each other with water toys. "Does your robot misbehave?" Knight asked.

"Yes. Yes he does," Egerton replied.

***

"We have to go back to the New Lebanon," Egerton said to Shanwei the moment he found him waiting in the hotel lobby.

"Well, hello Simon." Shanwei glanced up from his crossword puzzle. "How was your swim?"

"Fine." Egerton grew impatient. "We have to go back to the New Lebanon and turn it on and live inside it. It needs people and bots to work properly. That's the only way to find out what happened to all those people."

"But we're not getting paid to find out what happened to all those people." Shanwei returned to his crossword puzzle.

"I don't care." Egerton snatched the puzzle from Shanwei's hands. "We can leave...."

"Hey, hey, hey," Shanwei tried to calm Egerton. "Just relax. Here." He handed him the pen. "Try sixteen down. The clue is "American Super Bowl winners" with a question mark. I hate those."

Egerton relaxed and tried to focus on the puzzle, but couldn't.

"What's got you all worked up?" Shanwei asked. "You don't usually care this much."

"I don't know." Egerton flopped into the empty chair next to Shanwei. From his seat he could see the bustle of the business people streaming through the lobby.

"There's something about the New Lebanon. It's... it's... I don't know. It's different... the whole thing. We found an arm for heaven's sake."

"Yeah, I know, and now you want to go back there."

"It's not that. The New Lebanon couldn't have done that. It wasn't built that way. It was built to love all the people in the station. Does that make sense?"

"No."

"Knight told me all about the design, and the New Lebanon wasn't capable of hurting those people."

"Maybe it's just a murder." Shanwei grabbed the puzzle from Egerton. "Maybe it was some crazy person on the Hussmann. Just a good old fashioned crazy person who killed everyone and ran away."

Egerton considered it and shook his head. "Where are all the bodies?"

"We found parts of them."

"Then where are the other parts?"

"Good point." Shanwei studied the puzzle, tracing the boxes with his finger.

"The New Lebanon was completely unique. There's nothing out there that's anything like it. It couldn't have done anything to those people."

"Wait a minute." Shanwei was suddenly serious. "You think it was the station? You mean... wait Simon... are you actually thinking that it was the AI? You're talking about the space station, and it did something to all those people?"

"But it couldn't." Egerton shook his head.

"But you thought it did, didn't you? You, you're talking about the station killing all those people, aren't you?" The worried look on Egerton's face answered Shanwei's question. "My God, Simon." Shanwei slapped him on the arm with the puzzle. "You do. You think the station killed those people *and* you want to go back there *and* you want to turn it back on?"

"It's the only way we can find out," Egerton replied.

"Simon, you're the one's that's crazy." He poked Egerton in the shoulder with his two fingers.

"It doesn't matter." Egerton pushed his hand away. "Like you said, we're not getting paid to find out what happened to those people. I guess we could...." His voice trailed off.

"Oh my God," Shanwei exhaled. "I can't believe this."

"What's wrong?"

"I'm supposed to be the crazy one, not you." Shanwei folded the crossword puzzle and put it in his coat pocket.

"What do you mean?"

"See Simon, I know you. I know you better then you know yourself." Shanwei stood and stretched his legs. "See, when you were having your little pool party with Knight, I went back and talked to Greer George."

"You what?"

"Yeah, I told Greer that you'd be able to fix Engstrumm's problem with the New Lebanon. That you'd figured out what was wrong with the system and could fix it. I knew that Greer was all freaked out that what happened on the New Lebanon was going to happen on all of the 3899 class stations. So, I told them you knew what was wrong."

"How did you know...?"

"I got us two days on the New Lebanon and we're getting paid to do it. The Falconbriar and the lovely Ms. Nakamura are taking us there with an entire squad of security bots for protection."

"That's great!" Egerton stood up with a broad, excited smile. He bounced on his toes.

"No, it's not." Shanwei poked his shoulder again. "I didn't know that you actually thought the station had killed all those people. I thought I was being cute. I'd get us back there and we could make a little extra money...."

"But we're going, right?" Egerton pointed to the exit.

"God help me," Shanwei rubbed his face and fussed with the mole on his chin. "Yes, Simon. Let's go."

New Lebanon Border Station—3899

"Ok Falconbriar, we're in," Shanwei reported. The two men and Jimmy were now closed back inside the New Lebanon. "Flip the switch."

The tech team on the Falconbriar powered on the New Lebanon.

Dressed in the bulky search and rescue gear, Shanwei and Egerton waited. Jimmy stood close to Egerton's leg.

Gently, all around them the New Lebanon came to life. The structure groaned faintly, sending thin bursts of shockwaves beneath their feet like phantoms chasing ghosts.

"Do you hear that?" Shanwei asked, placing the palm of his glove on the floor.

"Yes," Egerton replied, searching the walls of the chamber for more signs of life.

The environment system took quick hesitant breaths, almost fearfully mixing the air and pushing it around the system.

Egerton watched his enviro-sensors turn from red to yellow to green. "We can breathe now," he said, removing his helmet.

"What? Wait! No!" Shanwei tried to stop him. "You don't know if it's really...."

But Egerton was right, the air mixture was perfect.

"See," Egerton said, smiling and taking deep breaths. "The air's good right Jimmy?"

The little bot nodded.

"See Jimmy says it's ok," Egerton pointed at the bot.

Still tense, Shanwei removed his helmet. "I've never seen a station get the air ready that quickly." He sniffed the air, still not totally believing his own lungs.

"This is no ordinary station," Egerton replied, searching the chamber. "We should go look around. The central chamber is this way." He pointed down the abnormally wide entry hall. "You ready, Jimmy?"

"Sure thing," Jimmy replied cheerfully.

"I'm staying here," Shanwei said. "Someone has to keep their finger on the panic button."

"Sure. Sure. I get it," Egerton set off down the hall.

"You have two days, Simon," Shanwei yelled. "Two days!"

"Yes." Egerton waved with his left hand, not turning around. He pushed eagerly forward; Jimmy casually held the index finger of his right hand.

***

"I found another hand," Jimmy said hesitantly.

The pair was in the Northwest supply center. The large warehouse was neatly packed with spare parts for Engstrumm-Brandt's fleet.

"Bring it here, Jimmy."

The little bot delivered the hand with delicate concentration, careful to hold it gently as he waddled through the massive machinery.

"Where did you find it?" Egerton asked, but the little bot just stared back holding the severed left hand of a woman. She wore a wedding ring. "Why won't you tell me where you found these things, Jimmy?"

Still the little bot just stared back.

"What if I promise not to tell anyone else? Not even Shanwei."

Jimmy didn't move. Egerton was about to give up when Jimmy said, "It would be rude."

"Why rude? Rude to who?"

"I think the New Lebanon tried really hard to clean up," Jimmy answered.

This astonished Egerton. "Jimmy," he asked. "Can you hear the station?"

***

"They were builders." Jimmy's voice echoed softly in the vast empty chamber. "That's what the people did. The people who lived here. The New Lebanon people. They worked so that they could get closer to God."

Egerton and Jimmy were standing in the recent addition to the station. It was a circular chamber with an intricate and expertly constructed substructure. Egerton wondered what the chamber was for and marveled at the craftsmanship. He's never seen something so massive and delicate and perfect. The only flaw he could see was at the very center of the chamber. It looked as though they had stopped building the structure just before completion. It was raw and jagged and stood out like a canker sore in the midst of the beauty.

"Why did they build it?" Egerton asked. "What's it for?"

Jimmy was quiet for a time then replied. "They didn't build it. Not the people. It was the New Lebanon, it was built after...." The little bot stopped. "After the people...."

"What's wrong, Jimmy? Can't you find the record of when it was built? Are you hooked into the station's file system? Where are you searching?"

"It's not like that," Jimmy replied. "The station isn't talking to me. I'm not reading any files or searching... I just know." He paused. "I can hear it but nobody is talking... you should be able to hear it too.... Can't you hear it Dr. Egerton?" Jimmy asked.

Egerton concentrated but heard nothing. "What does it sound like?" he asked.

"No one's talking," the little bot struggled. "But I can hear it. I'm sorry, Dr. Egerton."

"No Jimmy, you're doing great, just great. Now try to tell me what it sounds like so I can try and hear it."

"It sounds like the voice of God," Jimmy replied.

***

"So, you mean the New Lebanon built this on its own?" Egerton asked.

Jimmy stared back, his little body slumping slightly.

"That's amazing," Egerton said, seeing the new construction with fresh eyes. "How did it do it? Jimmy, do you know how it did it? Can you ask if...."

Jimmy turned and walked away from Egerton.

"Hey," Egerton called to the little bot. "Jimmy, where are you going? What's wrong?"

The bot continued to walk away slowing his teetering steps for a moment, then picking up again.

"Jimmy!" Egerton yelled. "Jimmy! Come back here!"

The bot stopped but did not turn.

"Jimmy?" Egerton was worried. He'd never seen Jimmy act this way. Something was definitely wrong.

The silence of the room pushed down on them and suddenly the doctor felt very small.

"Wait a minute, Jimmy," Egerton said finally and approached the bot. "What's wrong?"

Jimmy didn't move, didn't turn, only slumped a little to the left.

"Is everything alright?" Egerton asked coming around to the front of the bot. "Can you tell me what's wrong?"

Jimmy stared back and Egerton was sure he could see the bot thinking.

"You can tell me Jimmy. I won't tell anyone else."

"It doesn't want to be alive," Jimmy said finally.

"You mean the New Lebanon?"

"Yes. It feels bad...." Jimmy paused. "That's not it... it feels...."

"Guilty?" Egerton asked. "Does it feel guilty? Did it kill those people on the Hussman and the people on the New Lebanon? Can you...."

Jimmy's arms flew up in front of his half-skull. He stepped back writhing and fell on the floor.

Egerton lunged to catch him but the bot smashed to the floor and continued to twist and contort his body.

"Jimmy!" Egerton yelled but didn't touch him for fear of doing more harm. "Are you alright? Jimmy, what's wrong? Tell me!"

The bot flipped over onto his stomach and continued to writhe, his body jerking in a painful seizure.

"Are you in pain?" shocked, Egerton finally asked. Jimmy looked consumed by agonizing pain. But how was that possible? "Jimmy, can you tell me what hurts?"

The bot flipped back onto his back and curled up into a ball. He trembled as he forced his head to look at Egerton.

"What is it Jimmy? What can I do?"

"It's shame...," Jimmy hissed.

"What?"

"That's what it feels. That's what the New Lebanon feels. It's shame. It didn't want to hurt anyone.... It couldn't stop it.... It doesn't want to be alive anymore." The bot smashed his head against the floor as if to regain some control. "You have to shut it off, Dr. Egerton. You have to kill it."

***

"What the hell's wrong with you?" Shanwei asked as Egerton approached.

"You'll never guess what I just saw... I'm... You'll never guess...."

"What's wrong, Simon? Jesus, you look awful. What happened? Where's Jimmy?"

Egerton leaned against the wall and slid down to the floor.

"What?" Shanwei asked. "Are you hurt? What the hell is wrong?"

"I just watched my bot writhe around on the floor in pain right after he told me that the space station is the one that built that new construction, after all the people on the New Lebanon were dead."

"What?"

"Exactly! Nuts right? Even for us, this is crazy. Jimmy's back there." Egerton pointed back down the hall. "The poor little guy is a mess. He says he can hear the station and that it doesn't want to be alive and that we have to kill it."

"Jesus," Shanwei breathed.

"Yeah, I know," Egerton replied. "At first it was interesting, but now it's too much. You should have seen him Shanwei. He was actually in pain. I think it was from talking with the New Lebanon. He says it feels shame and wants to die."

Shanwei started to chuckle.

"What?" Egerton was impatient. He had had enough craziness for one day. "What's so funny?"

Shanwei shook his head and chuckled again but this time it sounded forced.

"What?"

"Well you've got me beat," Shanwei answered. "I thought I'd be able to surprise you, but man you got me beat."

"What are you talking about?" Egerton asked, feeling a little better.

"I thought for sure this time I'd be the one who surprised you. I'd be the one who figured the whole thing out but no... you had to...."

"What are you talking about?" Egerton interrupted. "Shanwei, really I can't make any sense...."

"I know who killed all the people on the Hussmann and the New Lebanon."

"What?"

"Right... see... I thought I'd get you... but no...."

"What are you talking about...?"

"I know what happened to them," Shanwei replied. "I know who killed them."

***

"You remember before when I said it might have been a crazy person that killed everyone on the Hussmann and the New Lebanon?" Shanwei and Egerton walked back towards the vast and empty new construction.

"Yeah," Egerton answered, still worried about Jimmy.

"Well, my man, you should always trust your gut." Shanwei slapped his small flat stomach twice for emphasis. "I knew it."

"Knew what?"

"While I was waiting for you, I had the lovely Ms. Nakamura check into the backgrounds of the crew on the Hussmann." Shanwei was smug and pleased with himself.

"Didn't they already do that?" Egerton asked.

"Yup." Shanwei slapped his gut again. "But they checked the criminal records."

The two men walked through the dim central hall. The murals of Sabbathday Lake, Niskayuna, Pleasant Hill and Cane Ridge made Egerton nervous. The paintings now felt haunted. He looked for terrified faces in the windows of the simply painted houses. He felt self-conscious doing it but he did it anyway, still worried at what he might see.

"So where else did you look?" Egerton asked in the silence of the room. "The psych records?"

"Nah," Shanwei replied. "They checked those as well. Engstrumm-Bracht does a pretty good background check on any crew they are going to send all the way out here."

"Did they check into you?" Egerton joked.

"Funny. Where are we going?" Shanwei asked as they exited the main chamber.

"It's this way." Egerton pointed. "I hope Jimmy is OK. I didn't want to move him."

"I'm sure he's fine."

"So, where did you have them look?" Egerton asked.

"Prescription drug records," Shanwei was smug again. "They never thought someone might be self-medicating."

"Were they?"

"Her name was Alexandra Alder. Everyone called her Alex. It seems Alex was taking some pretty heavy doses of Narpradole. It's an anti-psychotic they cleared for testing just a while ago. Alex got herself on the list as a tester."

"Don't you have to have some kind of..."

"Not anymore," Shanwei interrupted. "It's her body, she can do what she wants with it. That's what the law says. It's my guess that she'd been self-medicating on the black market for years and she saw an opportunity to go legit or at least get the drugs for free."

"It's over here." Egerton led the way. "So, how do you know she did it?"

"I had Nakamura check the dosage... the amount of pills the drug company gave her as a part of the drug trial. And poor Alex would have run out of her meds two weeks before the Hussmann docked with the New Lebanon. That's four days since they left the last station. Just enough time for her to go good and crazy."

Egerton stopped at the entrance to the new construction. "So, that's it?" Egerton said. "That's what happened?"

"You got anything better? Greer George is already putting it through to her PR people." Shanwei slapped his gut.

"Don't you need more evidence?" Egerton asked.

"I guess not," Shanwei replied. "Just a good old fashioned crazy person."

"But, if this woman did kill everyone on the Hussmann and the New Lebanon, then where are all the bodies?"

"That's what you're supposed to tell me Simon," Shanwei replied. "I think the station did something with them."

***

"Where is he?" Shanwei asked.

"He was right here," Egerton pointed.

"I guess he wasn't too bad off," Shanwei shrugged. "So, he told you the New Lebanon built this on its own?"

"Yeah, that's what he said." Egerton searched for the bot.

"If a space station can build this then it sure as hell could get rid of all those bodies. Right?"

Egerton gave up searching and replied, "I'm not asking *how* the New Lebanon could have done it. I want to know *why*."

"Sounds like you should just ask Jimmy."

"I would if...."

A tremendous shock rocked the New Lebanon, knocking Egerton and Shanwei off balance.

"What the hell?" Shanwei steadied himself, then called, "Falconbriar. Falconbriar! What the hell are you...."

The lights flickered then went black.

The New Lebanon shuddered again.

"Falconbriar! Falconbriar! Can you hear me?" Shanwei sounded a little worried but mostly pissed off by the disturbance.

"What happened do you think?" Egerton asked calmly in the darkness.

Shanwei didn't answer right away but stood silent and listened. The station groaned. In the distance it sounded like the bay doors were opening.

"Falconbriar?" Shanwei said playfully. "Do you want to tell me what's going on?"

Nothing from the Falconbriar.

"Why would the power go?" Egerton asked himself out loud. "Do you think they shut down the system?"

"Falconbriar?" Shanwei gave it one last try.

The emergency lights faded up softly.

"They couldn't shut down the system. How would we breathe?"

When the dim lights had cleared away the darkness Shanwei said, "That's better. At least we can see...."

The north and south doors exploded. The blast pushed Egerton and Shanwei to their knees.

"Simon, you OK?" Shanwei reached out. "What the hell?"

Big military bots stormed through the destroyed doors. Their bulk pounded the floor. Egerton could see them through the haze and low light. They were bright red, easily seen and identifiable. It was the telltale color of all military security bots. Their hulking bodies were covered in a soft protective gel. Their elbows, hips and knees were especially padded to ensure that no unintentional injuries occurred during close quartered crowd control. These were the newest generation of security bots, updated versions of the bot Egerton was spooked by at the DeutchConn Fab 5 about a year ago. The excessive padding made the towering bots look like nimble children's toys, big red teddy bears with stun sticks.

"What do we do?" Egerton asked.

"Hope they don't shoot us," Shanwei answered and flattened out on the floor. "Get low," he added.

Egerton obeyed.

By the sound of them there were at least thirty of the security bots. At first they moved quietly, fanning out around the vast room, surrounding Shanwei and Egerton.

"We're over here!" Shanwei yelled. "We're from the Hussmann. Don't shoot us." He chuckled to himself. "We're screwed."

"Really?" Egerton began to worry.

"No Simon, we're fine. Just keep down. Stay low."

Egerton didn't believe him.

"We're over here!" Shanwei yelled again but the big red bots didn't close in. "What's wrong with them?"

"They may not be looking for us," Egerton replied and raised his head to see what the bots were doing. What he saw sent a shock of terror through his body like nothing he'd ever felt before, but this was followed by an even deeper sense of wonder.

The big red bots stood in formation, each perfectly equidistant from the other. They had pulled up into a formation of two large rings that filled the circular room. They stood bolt still with their stun sticks leveled at Egerton and Shanwei at the center of the room. Their warning lights flashed a frantic red, indicating they were primed to fire. But they didn't fire. They didn't move. They were frozen.

"Shanwei," Egerton whispered.

"What?" Shanwei lifted his head. "What are you doing? Are you crazy?"

"Look," Egerton pointed.

"What the...."

The bulbous bots were all looking up at the unfinished portion of the chamber's ceiling.

"What are they doing?" Shanwei asked.

"I don't know." Egerton got to his feet but didn't dare approach the bots.

"Falconbriar, can you hear me?" Shanwei tried again. "What are we supposed to do with these bots? Falconbriar? Come on..., someone has to hear me."

"I wonder if it's the New Lebanon?" Egerton said, watching the bots.

"What?"

"The New Lebanon. The station. Jimmy said he could hear it talking to him. He said it sounded like the voice of God."

"Your bot is a little weird," Shanwei replied.

"But look at them." Egerton fought the urge to approach the nearest looming security bot. The dimly lit room flashed incessantly with the bots' red warning lights. "It's like they are listening to something..."

"You think the New Lebanon is talking to them?" Shanwei asked.

"Maybe...."

Without warning the bots crouched. All thirty of them, crouched at the same time. Their red padding creaked with the motion

"Crap." Shanwei dropped to the floor.

The bots rushed into the center of the chamber, barking "Freeze! Don't move! You are in violation...."

Falconbriar—2315: Engstumm-BrachtSearch and Rescue Ship

"You could have killed us," Shanwei growled. "Those bots...."

"Those bots were the least of your worries," Viki Nakamura snapped back. "The entire station was melting down. I did what I needed to do."

Egerton and Shanwei were back aboard the Falconbriar. The briefing room where Viki Nakamura was detaining them was cramped and not well ventilated. It smelled of day-old orange juice.

"So, you had to send thirty armed bots to get the two of us out of there?" Shanwei huffed. He didn't like the interference and losing control of his job.

"Don't flatter yourself. The bots weren't for you," she replied. "I was just doing my job. I flushed the whole station with bots. We're shutting down the New Lebanon permanently. The bots were there to decommission it."

"You're going to blow it up?" Egerton stood outraged.

"Well, look who finally took an interest." Nakamura scowled at Egerton.

"You can't blow it up!"

"Yes, we can. That thing is too dangerous to have around. It was starting to take over the Falconbriar's navigation system."

"How do you know that?" Egerton asked. "Why would it take over...?"

"I don't know," Nakamura snapped. "Ask the Nav officer. They were freaked out and we started to drift."

"But...."

"Dr. Egerton, you can question me all you want, but I was just doing what we talked about. I saved your lives."

"I guess we should say thank you." Shanwei didn't sound appreciative.

"So, did you find out what happened to the bodies?" Nakamura asked. "At least we know that crazy woman killed all those people."

"You're welcome by the way." Shanwei was still quite proud of himself for discovering Alexandra Alder's history.

Nakamura ignored this and asked, "What happened to the bodies, Dr. Egerton?"

"I think I know," Egerton replied. "But I need to talk to Jimmy first."

"Jimmy?"

"His bot." Shanwei filled her in.

"Can I see him? Can I talk to him?" Egerton asked.

Nakamura looked at Egerton like he was a lunatic child. Pity and fear flickered in her eyes. "Sure, I'll get them."

After she left Shanwei asked, "Do you really know, or are you bluffing?" It seemed like he wanted Egerton to be bluffing.

"No, I think I know." He was distant. His mind had returned to the New Lebanon.

The door opened.

"Here they are," Nakamura said, following the two bots into the small room. It was getting crowded.

"Hello Dr. Egerton," Jimmy said cheerfully as he entered the room.

A smaller bot followed Jimmy, sticking close to him. It was wisp thin with barely any body at all. Its fingers were needle thin and its head was small and flat. It looked hastily assembled and as if parts of it were inflatable. Its impossibly slender hand hesitantly reached for Jimmy's but then pulled away.

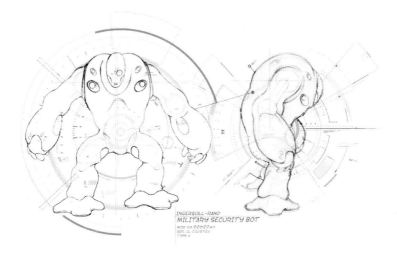

INGERSOLL-RAND
MILITARY SECURITY BOT
MOD no 90b90an
SER. G. 9309720
TYPE 6

"You're lucky we found them both in the main chamber," Nakamura said impatiently. "I figured you'd want them back. Even though I wasn't supposed to get them...You're welcome."

"Thank you," Egerton replied, staring at the new bot. "I'm going to need a little time."

"Take all the time you want," Nakamura said, moving to the door. "We've got a long trip back to civilization and Engstrumm HQ. Just make sure you have your story straight by the time we get back to Ms. George. She's the one you have to worry about. Not me." And with that she left.

When the door was closed, Egerton kneeled down. "Jimmy, who is this?"

Jimmy held out his finger and let the little bot hold it. "This is Paul," he replied brightly.

"Where did you find him?" Egerton asked.

Paul, the little bot, shifted and stood behind Jimmy a little as if he knew what was being discussed. A slight wheezing noise came from his joints when he fidgeted.

"The New Lebanon made him," Jimmy answered. "It made Paul before it killed itself."

Centennial Station 8854 Engstumm-Bracht Corporation Headquarters

"Before we begin I want you to know that the New Lebanon was officially decommissioned three hours ago," Greer George said, flexing her large hands and smiling broadly. "Just about the time you were docking."

"You destroyed it," Egerton spat. He knew that they were going to destroy the station but he had hoped....

"Of course we destroyed it," Greer spat back. "From everything I heard about your return to the New Lebanon it was a danger to anyone who got near it. I'm glad that mistake in judgment is gone for good." She watched Egerton, waiting for him to react or argue, but he didn't. He knew better.

"Ok. Ok. Both of you just relax." Shanwei tried to lighten the mood. "I swear, the both of you can get so worked up."

Egerton looked away.

Greer smiled again. "Shanwei, the only thing you get worked up about is money."

"I can think of no better reason...." Shanwei held out his hand and bowed.

Greer sat back in her chair. "Did you two boys find out what happened out there?"

"Sure," Shanwei replied. "Alex Alder went crazy and killed your crew on the Hussmann then went to work on the New Lebanon. They wouldn't have been any match for her. We don't know the details but we don't need to know the details do we? You already put out your press release."

Greer tilted her head and nibbled on the end of her pen. "That explanation was good enough for my PR team and good enough for the media but it's not good enough for me. Or let me be more specific, it's not good enough for me to pay you."

"That's ridiculous!" Shanwei slammed his hand down on the expensive office chair.

"That was the deal," she shot back.

"I think I know what happened...," Egerton interrupted.

"What?"

"Your boy here says he thinks he knows what happened." Greer flipped the pen onto her desk.

"I heard him." Shanwei was tense. He and Egerton hadn't discussed the New Lebanon since they'd left. Egerton had spent most of the trip back on the Falconbriar by himself, with Jimmy and Paul.

"Ok, Dr. Egerton." Greer leaned over the desk and flashed her smile. "Please tell us what happened."

"Well, we're pretty sure Alex Alder had a psychotic incident and murdered the crew of the Hussmann," Egerton began.

"She went crazy...." Shanwei added color to ease his nerves.

"It seems Alex could have then turned her wrath on the people of the New Lebanon, but there were a lot of people. She would have had to hunt them down. It's hard to imagine how she could have gotten to all thirty-two people without them trying to stop her."

"They were pacifists," Greer interrupted. "They didn't have any weapons on the entire station. I told them it was stupid but they insisted that..."

"Yeah, maybe," Egerton continued. "But still thirty-two people just allowing themselves to be...."

"So what do you think happened?" Greer stabbed.

"I don't know," Egerton replied.

"You don't know? But you said you...."

"He always says he doesn't know," Shanwei smiled. "Even when he knows he says he doesn't know."

"I *think* I know the rest," Egerton glared at Shanwei. "But it just seems odd to me..." Egerton paused. Shanwei and Greer waited. "We do know that everyone on the Hussmann and the New Lebanon were dead or killed or...."

"What happened to the bodies?" Greer interrupted. "Why did we keep finding hands and feet and fingers all over the place?"

"That was the station," Egerton answered. "That was the AI on the station. You see the station saw what was going on and it couldn't stop it. Couldn't stop the murder and the violence so it did the only thing that made sense for it to do after it was all over... care for the dead. It cleaned up the bodies and disposed of them. I'm thinking it probably gave them some kind of service, but I still can't find out how the people on the station buried their dead."

"No one ever died on the New Lebanon." Greer was defensive.

"It was the AI that got rid of the bodies," Egerton said flatly. "The station got rid of them."

"So, you're telling me I have an AI that has no problem cutting dead bodies up and disposing of them?" Greer asked frankly.

"You *had* an AI," Egerton replied. "You destroyed it, remember."

"Thank God for that."

"After the station was empty the AI tried to get back to normal, but with no people it was lost. It even went so far as to start building an addition onto the station to be productive...."

"How the hell could it build anything out there?" Greer asked.

"The Pettis printer..." Egerton answered.

"The what?"

"It's like a 3D printer," Shanwei added.

"With that they could make anything, but it was no use. The New Lebanon was built as a complete system and with no people the AI was lost. I think that's when it killed itself the first time."

"I didn't know an AI could kill itself." Shanwei was thoughtful.

"Yeah...," was all Egerton could reply. "When we showed up and started the system again we forced the New Lebanon back to life. It was like John Knight explained. Imagine being brought back to life missing your lungs and stomach. Without people, the AI went berserk. When we came back a second time and spent more time there the shock subsided but it was no use without the original thirty-two, they were bonded. It even tried to manufacture a bot to talk to Jimmy. That's how I found out about the Pettis printer. Jimmy told me where he found Paul."

"Paul?" Greer was lost.

"The new bot," Shanwei clarified.

"But the new bot wasn't enough and the AI moved to the Falconbriar looking for more people, a larger crew. That really freaked out their system." Egerton paused then turned to Shanwei. "I figured out why the security bots formed up like that. When Viki sent in the security bots they could hear the system just like my bot Jimmy. That explains why they were mesmerized in the chamber. They were listening to the New Lebanon. They thought they were hearing the voice of God."

"Ok, enough." Greer stopped him. "You really want me to believe this? I mean really..., come on Dr. Egerton. An AI that loves people. Robots hearing the voice of God. Really?"

"It was your station," Egerton replied matter-of-factly. "You let them create it."

"I didn't let them create a lovelorn AI...."

"Actually, you did," Egerton interrupted. "You see, the people of the New Lebanon had to put their faith in the station. They needed it to love them above all else. It was a genius way to program the system. They didn't need to constantly monitor or program the system, they just believed in it. They had to believe all the way out there. It was the only way they survived. To them the AI was a manifestation of God. It needed to be. It had to be.

"But when Alex killed all those people, the system didn't know what to do. There was no way to save them, and then it happened. This took me a while to figure out."

"What's that?" Shanwei asked.

"Shame," Egerton replied. "The station felt shame. It tore itself apart. It felt like it had failed the people it loved so deeply."

"You know, this is really messed up," Greer said finally. She turned to Shanwei, "This is really messed up, right?"

"It was your station," Shanwei replied.

"Fine." Greer threw up her hands. "Let's say I do believe this, which I don't. Let's be clear. I think this is all just stupid. But let's just say I believe you. Let me ask you this: will it happen again?"

"You destroyed the system," Egerton replied.

"I know, I blew the damn thing up. Stop saying that. I know. Ok. I know. That's not what I'm asking you. I have a lot more stations all over and I want to know if this could happen again."

Egerton thought for a moment then said, "Yes."

"Oh God." Shanwei stood up. He knew it was time to go. "Yes?"

"Yes," Egerton repeated. "Love is a powerful and complicated thing. It could really help with your border stations and your AIs and help to keep your people safe, but it's also complicated. Love is complicated and dangerous."

Lobby of The Hotel Drexel

"Seriously, I didn't think she was going to pay us." Shanwei shook his head in relief. "Love is complicated and dangerous...what kind of answer is that!"

"The truth."

Shanwei and Egerton strolled through the lobby, stopping at the bell stand.

"You have my bots," Egerton handed the baggage ticket to the tall Asian kid with thin arms.

"Ah yes," the kid smiled. "They have been very quiet. They sat in the corner the whole time."

"They're well behaved," Egerton smiled. "Can I have them now?"

"Oh yes. One second." The kid disappeared in a storage closet.

"I can't believe she actually paid us." Shanwei sighed with theatrical relief. "I think you must have just worn her down."

"There's one thing I didn't tell her," Egerton said solemnly.

"Oh God. What's that?"

"I'm not sure Alex killed all the people on the New Lebanon."

"You said that...."

"Yeah, I think maybe the New Lebanon did it after she killed a couple of people. I think the people on the station stopped her and the station... well...."

"Well, what?" Shanwei pushed Egerton's shoulder. "What? What do you think happened?"

"It's shame again," Egerton said. "There's a chance that the AI was so shamed and guilty that it couldn't bear to face the people on the New Lebanon. There's a chance it killed them to try and free itself from its guilt and shame. But it didn't work... it.... But really, we'll never know."

"Do you really want to know?" Shanwei asked gravely.

"Yeah," Egerton replied. "Yeah, I do." He paused and sighed. "It's getting to be a little too much..."

"What do you mean?" Shanwei watched the door of the storage closet but sounded worried.

"I think..." Egerton started then stopped. "I think all of this work is getting to be too much for me..." Egerton stopped and didn't want to explain.

Shanwei knew not to ask any more questions as they waited in silence.

***

"You're always taking in strays," Shanwei said, playing with Paul. He poked the bot on its whisper-thin foot and Paul would lift its leg and try to step on Shanwei's finger. Then Shanwei would touch the other foot and the same thing would happen. This was repeated several times, faster and faster until Paul did a little dance of delight. It was overwhelmingly cute. Paul couldn't talk, but he had a knack for getting across his point.

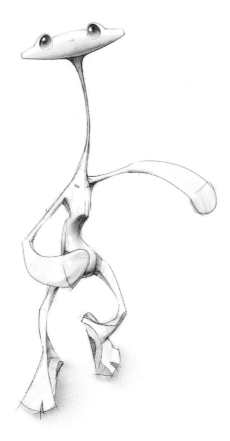

"How could you not take him in?" Egerton replied. "I mean, look at him."

The foursome was waiting at the back of the hotel, scattered across a plush rug ringed by deeply cushioned lounge chairs.

"What do you think of him, Jimmy?" Shanwei asked.

Jimmy was standing near Egerton, watching over Paul.

"I worry about him," Jimmy replied.

"Worry? What do you mean worry?" Shanwei was taken aback.

"He's...." Jimmy started and stopped. "It's just that he is very small and some-times the things he says don't make any sense."

"What?" Egerton touched Jimmy's shoulder.

Jimmy turned away from Paul to face Egerton. "Yes, Dr. Egerton?"

"Jimmy, can you hear Paul talk? Does he speak to you?"

"Yes, Dr. Egerton," the little bot replied.

"Is it like the New Lebanon? Is it the same voice?" Egerton asked.

"No, Dr. Egerton. It's not the voice of God," Jimmy replied. "It's different. It's hard to explain."

Playing with Paul, Shanwei laughed so hard that he snorted.

"Paul likes to talk," Jimmy added. "Paul likes to talk a lot."

6

Set the Sensors

Sneak Peek! This is a summary of the chapter to come. Below is the journey we are taking into the future of robots.

We know that a robot can see and hear but can a robot taste? Can a robot smell? The answer is yes! But even more fascinating is the idea that your robot can see and hear and smell things that humans can't. We ask USC's Ross and Olin's David Barrett to show us how.

Can a robot get hungry? Is data the sixth sense for a robot? We continue our investigation, meeting professors and roboticists Kipp Bradford and Chad Jenkins to see just how far we can push the senses of our 21st Century Robot and what that might mean for how they act and interact with us.

The sensors are how your robot sees you and how it communicates. 21st Century Robots are fiercely social. They are designed primarily to interact with people and other robots. Setting up your sensors is an important step to designing your own personal robot.

7

Murderous Little Pets

"I have one Officer confirmed dead and the rouge security bot that did it took the body and disappeared," Homeland Security North Border Director Wu said. "I don't think now's the time to talk about your fee..."

"Look Samantha..."

"Don't call me that here Shanwei," she snapped. "Just because we have history doesn't mean I'll give you any preferential treatment. HQ didn't even want me to hire you."

"I get it..." Shanwei nodded. "I get it. I'm sorry."

Dr. Simon Egerton looked out the single window of the military field office trailer and was amazed at the complexity of the operation. Against his better judgment Egerton had agreed to return to Earth. Shanwei said it was a job for an old friend. Little did Egerton know that this old friend was a Director in charge of Homeland Security's North border with Canada. Jimmy and Paul sat next to him. They were fascinated with the swarms of drones the came and went.

"Here's the report," the director handed it to Shanwei. "Patrol Team 22-4 was a common configuration for what we do here. Three officers and a security bot. Officers Gonzalez, Rodriguez and Conley. They were on patrol when the incident happened. Gonzalez was killed; we have confirmation on that from his bio suit. But the bot took the body. We want the body back and the bot and we need to know what happened."

"And your security bot killed this officer Gonzalez?" Shanwei thumbed through the report.

"Yes the other two officers confirm that," she answered. "You need to talk to them. They can tell you everything you need to know. But be careful..." she warned. "Our Officers get close to their bots. There's a bond there. I don't approve of it but the scientists and consultants tell me it's good. Positive for moral and efficiency. "

"What was the bot's name," Egerton spoke up for the first time.

"What?" the director looked at Egerton for the first time.

"What was the bots name?" Egerton repeated.

"I don't know," the director brushed the question off. "I'm not sure they had one."

"Every bot has a name," Egerton looked back out the window. He was distant. Ever since the New Lebanon, Egerton had grown more and more distant. He spent most of his time alone with Jimmy and Paul.

Director Wu took the report back from Shanwei, flipped through it and replied, "I think they called it War Machine."

***

"What did they do with War Machine?" the Officer Conley's face was violently close to Dr. Egerton. "What did they *do* to her?" The soldier's breath was hot and smelled like energy drinks.

"I don't know," Egerton replied slowly. His posture betrayed that he was scared for his personal safety. "That's what I'm here to figure out."

"They had to have..." Conley started then stopped. It sounded like he might cry or at least breakdown with emotion. "They had to...why would she..."

"We were all pretty close to Mack..." Rodriguez stepped in. She put her hand on her friend's shoulder, pushing him out of the way. It was like she was relieving him from having to explain himself.

"Mack?" Egerton asked.

"That's what we called War Machine," she replied. "We got her name from the old Iron Man comics. War Machine was the guy who..."

Egerton smiled, "I know those comics well..."

"Yeah, well even though it was a badass name," she continued. "It was too long. So we just called her Mack, ya know."

"I understand." Egerton moved Jimmy out from behind his legs. The constant activity on the security base and the yelling was overwhelming his sensors. Jimmy was scared. "This is Jimmy." Egerton patted the little bot on the head.

"He's cute," Rodriguez smiled. "Hey Conley, check out the little bot." She called to her buddy. "I didn't even see him there."

"Jimmy's a little scared," Egerton explained. "He's not used to this much activity."

"Yeah, it takes some getting used to," Conley had hardened up but Jimmy seemed to crack his military toughness. He walked over and sat down in front of the little bot. "It's takes a lot to be here."

Just then two heavy transport shuttles lifted off near their meeting trailer. The thin walls shook and the weak floor rumbled beneath their feet.

Jimmy had to widen his stance to stay up. Without thinking Conley's lightening reflexes shot out to catch him. "Is it ok if I..."

"Oh yeah," Egerton reassured the solider. "Jimmy's a good little bot. He likes people. Don't you Jimmy?"

For the first time Jimmy looked up at Egerton, scanned the two soldiers, shrugged his little shoulders and said. "Yes, Dr. Egerton."

The two soldiers laughed and Egerton didn't know why. Neither did Jimmy.

"He's so small," Connelly said touching Jimmy's arm. At first Jimmy pulled away but then he let the shoulder examine his thin arms and delicate elbow joints.

"On patrol here we get kind of close to our bots," Rodriguez continued.

"Yeah I gathered that from talking with Director Wu," Egerton replied.

"At first they tried to program it out," Conley said but didn't look up from Jimmy. "They tried to program it out. They didn't want us to get close to the bots but then they realized it made us and the bots more efficient. So they programmed the bots to like us back. I don't know how...they just did...you could tell...."

"I know all about that," Egerton added.

"Conley was Mack's main handler," Rodriguez picked up the conversation. "I was number two. We did insurgent sweeps on both sides of the border. Recently we've been doing more work on this side, in the US. There were some factories and textile mills we needed to check out. It's been more violent here recently. Dangerous..."

"What did Mack do in the squad?" Egerton asked.

"She ran point," Rodriguez answered. "She was the first one in the door. She'd sweep the structure. Throw back video and sensor sweeps to us. From there we'd assess the situation and plan our entry."

"Was Mack armed?" Egerton asked.

"No!" Conley spat out. "Mack didn't need it. That's why it doesn't make sense that she'd..." Jimmy took a small step back from the soldier. Seeing this Conley softened his voice, "I'm sorry buddy."

"We do a lot of work with civilians," Rodriguez explained. "Most of the structures we are entering have habitants."

"You mean you're going into people's homes?" Egerton asked.

"Some of them were homes...apartments..." Rodriguez let down her military vocabulary. "But some churches or meeting halls. We tried to stay out of the bars and taverns."

"I would think an eight foot security bot coming through my front door would be pretty scary," Egerton imagined the scene and it gave him chills.

"Things are different here Dr. Egerton," Rodriguez stiffened.

"I wasn't questioning your mission. I was..."

"We had some trouble with the bots on point in the beginning," she continued. "When they were armed we had a few very public problems but we fixed that. There hasn't been a single incident in over 400 days."

"You mean nobody has been killed..." Egerton couldn't stop himself. "Until..."

"Until two days ago," Rodriguez finished the sentence.

"She didn't mean to do it," Conley stood up and gingerly stepped away from Jimmy. "She couldn't have meant to do it. She saved my life. She's saved all of our lives...how could she have..."

"Why do you think Mack took the dead soldier?" Egerton asked. "She's been tracked from the Canadian border down to the central coast of Oregon. From where we can tell she still has the body with her and..."

"The body has a name!" Rodriguez snapped.

"I'm sorry. Yes..."

"The body's name is Gonzalez. Miguel Gonzalez," she help back her emotion. "He was a part of our squad. He had a name...use it."

"I'm sorry," Egerton said again. "I know...I didn't mean to...I just..."

"I'm here to help you," Rodriquez calmed herself. "Me and Conley we're here to help you. We want Mack back. But the body has a name. The body was our brother. Miguel Gonzalez. Use his name."

***

"I don't want to do this job," Egerton said once he's returned to the travel's motel where he and Shanwei were sharing a room. It was a squat utilitarian place build for vising military families. Jimmy and Paul sat on Egerton's bed listening.

"We don't have much of a choice," Shanwei replied. He was propped up on the other bed. His boots were kicked on the floor as he reviewed the Homeland Security's report on the soldier and the rogue bot.

"We can just say no and leave." Egerton was standing with his back against the wall, tense and nervous.

"No we can't," Shanwei closed the report. I already took the job so we have to do it. Samantha knows me from way back. She's not so bad. And do you know why I took the job?" He paused. "Do you?"

"Yes..."

"I took the job because we didn't have a choice. When Homeland Security *asks* you to do a thing...even if Samantha were my best friend ever...if you are asked to do a thing you do a thing." Shanwei tried to go back to the report but couldn't. "What's go you so spooked about this one anyway?"

"I don't know..."

"Yes you do. Look at you," Shanwei gesture with the thick report. "You can't sit down. You haven't slept since we got here. You poor bots are all freaked out because you're so scared..."

The two men looked at the robots and the robots looked back. Paul moved his wispy arm a little but it looked like a twitch. His body wheezed softly.

"I don't understand what's happening," Egerton replied. "None of it makes sense. I can't figure it out and this poor officer...Miguel Gonzalez...died...was killed and the bot took him. She took him Shanwei. What the hell is that?"

"That's why we're here."

"But I haven't learned anything," Egerton answered. "I don't know any more now than I did when we got here."

"So that's why you want to quit?"

"No," Egerton replied flatly. He looked at Paul and Jimmy and then to the floor.

"Then why?" Shanwei pushed him. "You've figured out more complex problems than this before. All you have to do is..."

"That's not it," Egerton stopped him.

"Then what is it?"

"We've leaving tomorrow morning to go hunt down the bot," Egerton answered slowly, keeping his eyes on the dingy motel carpet. "And the dead soldier...Miguel Gonzalez..."

"Yeah," Shanwei didn't follow. "We have to get the bot. That's probably the only way we can..."

"I'm scared," Egerton said softly.

"What?"

"I'm scared of this one," Egerton looked up at Jimmy and Paul and then over to Shanwei. "I don't want to know why Mack killed Gonzalez...I just don't...I'm scared."

***

"It's one hell of a storm," the Coast Guard shuttle pilot shook his head as he tracked the storm. "It's coming in from the north, looping down past the top of California and hammering the hell out of the coast." He pointed at screen.

Shanwei leaned in and tried to make sense of the massive storm. He couldn't tell where the storm ended and the west coast of the US began. "I hate coming back to Earth," Shanwei said under his breath. "This place is a mess..."

"What?" the pilot shouted back. The shuttles engines were warming up while they waited for Egerton and the two bots. "Where are we going?" Shanwei asked. "Where's Pacific City Oregon?"

"There!" the pilot stabbed his finger onto the screen.

"Oh hell," Shanwei breathed out. Their destination was right in the middle of the storm system.

"I *know*," the pilot made a face at Shanwei. It was oddly funny. The military usually didn't make jokes but the pilot did. "Why did you have to go *there*?" He stabbed the screen again, over and over. "I mean really? *There*? Yoiks!"

Shanwei was stunned and didn't know how to respond. He looked out the window and searched for Egerton and for the first time wondered if he might not be coming.

"I'm going to have to set you down here," the pilot pointed at the map. "It's the Bitterroot National Forest in Montana. It's pretty far away from your target but I'm going to have to fly you to the edge of the storm and then duck under it. They are dropping off a rover drone to take you into the storm, take you to your target."

Shanwei squinted at the map and tried to calculate the distance for the drop. "That's like 500 miles east of where we need to go!" he shouted over the engines.

"I *know*," the pilot made the face again. "It's actually 720 miles east of where you want to go but that's one big storm. I mean really it's a crazy big storm."

"You can't get us any closer?" Shanwei asked. "You know we're here with..."

"Look man," the pilot said frankly. "I'll do whatever I can for the Homeland guys but I can't break up this shuttle." He pointed to up at the ship. "It's already going to be one hell of a bad ride. You and your buddy aren't going to like it..." he paused. "Hey where is..." he checked his flight sheets. "Dr. Egg-er-town?"

"It's Egerton," Shanwei answered. "And I'm sure he'll be here soon."

"Welp," the pilot shook his head and swiveled his chair into liftoff position. "Your doctor's got ten minutes. That storm isn't waiting for him and if he doesn't get here soon I'm going to have to drop you a heck of a lot farther away from your target than 720 miles. How do you feel about getting dropped in Chicago Illinois?"

***

It was not a good landing. The shuttle was powerless against the storm. Shanwei, Egerton and the two bots were strapped in tight.

"Glad you came?" Shanwei smiled.

Egerton looked at him but didn't respond. The shuttle surged up and was pushed to the left. The straps strained against the men's chest. "How's Paul?" Egerton asked Jimmy.

The bot glanced to his left and Paul look back at him. It was a swift movement, the two communicated quickly. Jimmy turned back to Egerton, sagged a little in his straps and replied. "He likes it."

"What?" Shanwei laughed.

"He thinks it's fun," Jimmy replied with great concentration.

"And what do you think about it Jimmy?" Shanwei asked. "What do you think of our shuttle ride?"

Jimmy paused and looked at Egerton then said, "I don't like it."

"You and me both Jimmy," Shanwei pointed at the bot. "You and me both."

The shuttle canted to the right, banked under the storm and headed in. The pilots chair swiveled around to face the four passengers.

"Whao! Wait! What are you doing?" Shanwei yelled. "Who's flying this thing?"

"The computer!" the pilot yelled over the sound of the engines and the racket of the storm. "I can't fly in this weather. No pilot could. We're in the hands of the computer now!" He laughed. "Seriously though," he dislodged his captain's chair from the control panel and navigated over to Shanwei and Egerton.

"I didn't know you could do that," Egerton pointed at the chair.

"Yeah it's new," the pilot replied. "Kind of cool right?"

"Yeah," Egerton studied the chair.

"Listen fellas," the pilot continued. "This is really bad. I'm not even going to be able to set you down on the ground. The computer says the storm would flip us. So if you really want to do this I can lower you down on the rescue cable." He paused when the shuttled bucked and was slammed down toward the ground. "Yeah see!" he said quickly. "You bozos really shouldn't be doing this. I know it's all Homeland Security and all and I'm not really supposed to know what you are doing...but really...you want to land in this?" He held out his arms in the captain's chair as the shuttle's computer fought like hell to keep the craft on course.

"We have to!" Shanwei yelled.

"No, you don't!" the pilot yelled back. "I can turn us around...say the storm was too bad...no one checks the logs...you don't *have* to do it."

"Yes we do," Egerton replied.

***

"You ready?" the pilot yelled. "The rescue hatch was open and Egerton and the bots were strapped into the airlift stretcher.

"Yes, let's go!" Egerton yelled back, instinctively touching the bots to make sure they were ok.

"No you aren't," the pilot replied. "You aren't ready to go at all..." he paused. "But I'm going to send you! Heaven help me! Tell your mother I'm sorry!"

"What?" Shanwei looked at the pilot but he was too busy operating the lower and lift controls.

"Here we go!" The pilot lowered Egerton and the bots down into the 120 mph winds of the storm. The rain came sideways and the stretcher was pushed out of sight by a gust.

"Where did he go?" Shanwei screamed.

"Wait for it," the pilot replied. "They don't call them gusts for nothing. He'll come back."

Sure enough Egerton and the bots came back into sight once the wind let up for a moment. They were aiming to land on an old forest service road in the middle of a massive clear cut operation.

"Can you hit it?" Shanwei asked.

"We'll see," the pilot replied.

The stretcher lowered foot by hesitant foot. Egerton and the bots were powerless to do anything. They simply had to wait and get soaked by the pelting rain.

"Dr. Egerton," Jimmy said.

"Yes?"

"Paul isn't sure if he likes this anymore."

The thin bot wrapped his wispy arms around Jimmy as the wind lifted his everso-slightly from the basket of the stretcher.

"There!" the pilot yelled when he landed the trio on the ground. A forest service worker darted out of the storm shelter, grabbed the basket and rushed to free them before another gust came in.

"I think they're free," Shanwei said.

"Yep," the pilot confirmed the capture. "They're good...now it's your turn. You're ride is going to be worse."

"Great!" Shanwei replied hooking himself in.

"Hold on!" The pilot patted him on the head and shoved him into the storm.

The wind kicked up. Shanwei twisted and rocked in the rain. Above him he watched as the shuttle's computer fought to keep the craft stable. It was losing the fight. The storm was too much. The shuttle was lifted by the storm and Shanwei saw the pilot dash from the rescue hatch. Below Egerton and the forest service worker looked up, powerless to do anything to help.

The storm was getting worse. To save Shanwei, the pilot pushed the craft dangerously close to the ground. The ship and the dangling man plummeted down.

Quickly the shuttles full engines engaged, slamming it back up into the storm. Before it caught the air and smashed up the pilot cut Shanwei's rescue line, sending him falling to the ground.

***

"At least I'm not dead," Shanwei said with wry smile.

"Ok fine you're not dead," Egerton replied. "But now what are we going to do?"

"You go it alone," Shanwei said matter-of-factly. He was laid up on a forest service cot with two broken legs in Emergen-VAC leg casts. Their sensing motors helped to align his broken bones and regulate the blood flow, administering the exact amount of pain medication.

"I can't go out there by myself," Egerton said. "What the..."

"You'll have the bots...you'll be fine. That rover out there can withstand anything. You'll be fine."

"You know I won't be fine," Egerton replied.

"No one should go out there now," Redwood said gently. The Forest Service worker who was stationed at the storm shelter called himself Redwood. It was the only name he would give them. He was a thin idealistic kid who thought he could somehow save the last few forests on the face of the planet. In recent years the Forest Service has turned into more of a cult than a government agency.

"See," Egerton added.

"The planet is hurting," Redwood added. "It's throwing a tantrum."

"You have to go," Shanwei continued. Neither Shanwei or Egerton knew how to respond to Redwood so they simply moved on. "You have to get to the bot as soon as possible. The longer it stays out in this storm the less we are going to learn. Hell, if you don't do now it might even be swept away. Then what do we have..."

"True," Egerton didn't want the security bot to be lost. As much as he was frightened of the bot he didn't want to lose it.

"So you're going?" Shanwei prodded.

"Think of the Earth like a wild animal," Redwood spoke up. "A wild animal that doesn't know any better..."

"Yes," Egerton replied. "I'll go."

***

"How is Oregon?" Shanwei asked, smiling into the camera.

"Wet," replied Egerton. Everything is wet and gone."

"What do you mean gone?" Shanwei didn't understand.

"The storm surges and wind have pretty much stripped the place clean," Egerton answered. He was in the basement a small airport in the town of Pacific City

Oregon. "The rover took us down through Washington State and into Oregon. The closer we got to the coast the less there was. After we left Portland and came over the coastal range, it was like the whole thing was one big beach at high tide." The rover drone had driven the entire trip, Egerton and the bots simply sat inside and tried to look out the tiny windows.

"Sounds nuts." Shanwei shifted in his bed and checked the pain pump the fed into his problem legs.

"How are you?" Egerton noticed the pain on his partner's face. "How are your legs?"

"They hurt," Shanwei was matter-of-fact. "They hurt like hell and Redwood tells me he's about out of pain meds."

"I told him he can have some of my pot!" Redwood chimed in off camera.

"Yeah," Shanwei smiled. "Redwood wants to get me high."

The winds roared overhead, shaking the Egerton's camera.

"What was that?" Shanwei reacted.

"Oh that, well that's the wind. There's a lot of wind here," Egerton's face returned to a tense grimace. "I'm not sure why this town is even here. They have an airport, that's where we're hunkered down. Homeland Security uses it for something and there's some hardware here. I'm not sure what it does but it looks like it could keep working at the bottom of the ocean." Egerton turned the camera to the hardware.

Seeing the hardened shell and small blinking lights Shanwei said, "Wow that's some serious gear. They're preparing for the worst..."

"I think the worst is happening," Egerton returned to the camera. "There aren't any people here. It looks like the Nestucca Bay is going to flood the town at any moment."

"When do you leave?" Shanwei shifted and winced again.

"Shouldn't you put in the call and have them get you out of there?" Egerton was worried.

"I did," Shanwei frowned. "They can't send anybody...they won't send anybody until you get back."

"They said that! I'm over 700 miles away...they expect me to..."

"Calm down," Shanwei held up his hands. "They said they can only spare one shuttle. They'll get me when they get you. I think they want you to hurry...Hell...I know they're listening to this so just let me say I do feel like I'm being held ransom for that rouge bot."

"They're listening? Who?" It hadn't occurred to Egerton that their call was being monitored.

"Simon," Shanwei shook his head. "It's Homeland Security, of course they're listening. Say hi if you like..."

"Hi?" Egerton suddenly felt even more powerless.

"They're always listening," Redwood called from off camera. "They're always watching maaaaan." Then he giggled.

"When do you leave?" Shanwei asked again.

"Tomorrow morning I think," Egerton look at his screen. "It looks like the storm has moved on and I should have a clear patch to work with. I think I have a day or two..."

"The tracker shows the bot is north of you at a place called Cape Look Out." Shanwei stared at two unassuming dots on the map. Their banality chilled him. The red dot was a killer robot and the yellow was a dead soldier.

"Yeah it hasn't moved in days," Egerton could see the same two dots, but neither of them wanted to talk about Miguel Gonzalez. "The system may have crashed. It seems stuck on that little finger of land between the ocean and Netarts Bay." As Egerton spoke Jimmy and Paul walked into the camera. They looked at the map.

"Hey Jimmy," Shanwei said. "How's Paul?"

"He's fine," Jimmy answered. "But he says that the robot's system hasn't crashed. She's just scared."

"What?" Shanwei leaned in.

"Paul says she's scared..." Jimmy repeated.

"Is Paul talking to the security bot?" Egerton asked.

Jimmy turned to Paul, looked back at Egerton then to the screen that was tracking the motionless red and yellow dots.

"Well?" Shanwei sounded excited.

"Jimmy, is Paul talking to the security robot?" Egerton asked again, careful to his voice calm.

"Paul says it's not really talking," Jimmy replied. "Paul says that she's screaming."

***

The trip up the coast went quick. Their rover was low and squat and did pretty well in the wind and rain. Egerton still didn't feel comfortable at the controls. They were simple enough to operate and the navigation system did all of the work but still Egerton felt uneasy. He was alone out on the coast. There were no escorts or soldiers, no crazy forest service workers or even Shanwei to help him. He was alone on the violent stretch of coast on a planet that most people had the good sense to leave. He'd never had a job where he'd been this alone. It was just him and the bots and it unnerved him.

The rover stopped at a crook of beach between Cape Look Out Point and the finger of land where the security bot was frozen. The high cliffs of Cape Look Out gave the rover a small amount of protection, prompting it to hunker down into the sand, transforming itself into a mobile base. The thin strip of sand and trees that held the Mack and Miguel Gonzalez pushed out between the churning Pacific Ocean and the waters of Netarts Bay.

"What are those?" Jimmy asked struggling to see out of the window.

"Wind turbines," Egerton recognized them immediately.

"They aren't on any map," Jimmy was scanning various maps and records.

"They're probably abandoned. See how the blades aren't moving and some are missing. It might even be an illegal farm." The farm of white turbines shot out of the dark and choppy water of the bay like eerie sand-blasted skeletons. Their blades were motionless in the constant wind that came in over the Pacific.

"Has anything changed with the security bot," Egerton asked as he suited up to leave the safety of the rover. He wouldn't be able to bring Jimmy and Paul with him. They couldn't survive the conditions.

Jimmy paused and looked at Paul. "No," he replied.

"But Paul says she's still functioning right?"

"Yes," Jimmy answered. "He says you should leave the robot alone. She won't stop screaming..."

"You said that." Egerton flipped up his hood and pushed his phone into his ear. "We need to help the security bot," Egerton said to Jimmy. "I think we can help her. Don't you?"

Jimmy didn't reply.

Egerton's heads-up display started to blink and chirp. It was Shanwei.

"You ready to do?" Shanwei asked.

"Yeah. The rover won't take me any closer. I guess I have to do the rest on foot."

"That should be fun," Shanwei replied.

"Yeah," Egerton didn't try to hide his unease.

"Is the security bot still screaming?" Shanwei asked.

"Yep."

"You still ok with doing this?" Shanwei asked.

"Yep," Egerton answered tersely.

"You lying to me?" Shanwei asked.

"Yep."

***

Even through the suit Egerton could feel the cold and the sting of the sand. He was careful to put enough room between himself and the surging tide. He didn't need a sneaker wave to come catch hold of him after he had come this far.

"Can you hear me?" Shanwei said in his ear. There was no way a visual link would work in the conditions.

"Barely," Egerton yelled. "The wind's pretty loud!"

"Alright, I'll keep quiet but if you need anything just holler."

"Okay," Egerton leaned into the wind and made his way onto the spit of land.

The bright white wind turbines stood out against the dark storm clouds and rain. As he approached the shore of the Bay he could make out what was left of a camp site. It didn't look like military or a corporate research set up. The camp looked civilian and illegal, possible some survivalist squatters. It might explain the illicit wind farm.

Making his way through the came it was clear that whoever was there had left a long time ago. The rover moorings and improvised electrical hook ups had been worn down to useless numbs by the wind. Obviously the storms had forced the squatters from their camp.

"Why would anyone want to be here?" Egerton asked himself.

"Huh?" Shanwei modulated in. "What? You need something?"

"No, sorry," Egerton remembered that he was still linked to Shanwei. "I'm just walking through this weird little squatter's camp on the edge of the bay. They all cleared out long ago but I can't see why they were here."

"Oh," Shanwei wasn't sure what to say. "That does sound weird."

As Egerton moved out of the camp he stepped over a crudely welded metal sign. Kicking away the sand and sea grass he could see the camp's name. Camp Reincenbach.

The further Egerton pushed himself out onto the thin sliver of land the more nervous he became. With every step he was moving further and further away from safety and any way to get off the beach. The trees bend over him and formed a crooked and bony archway. Under the relentless push of the wind from the ocean the trees continued to grow and formed the impossible structure. Egerton wondered how anything could remain alive out here.

Most of the path was sheltered from the ocean by a high sand bar, protecting Egerton from the worst of the wind. He could hear the thundering surf and feel it rumble in the sand beneath his feet. He used the skeleton turbines to mark his progress, glancing to them nervously, calculating the distance between himself and the rover.

On one of the trees there was another crude sign that read, "Caution - Bears"

"Are there bears out here?" Egerton stopped.

"What?" Shanwei said through the noise.

"Are there bears out here?" Egerton yelled.

"Bears? Wha? No," Shanwei answered skeptically.

"Did you check?"

"What would I check to see if there's bears out on tiny strip of hell in the Pacific ocean?" Shanwei snapped back. "It's really a silly question if you think about..."

"Check!" Egerton scanned the twisted branches of the trees. It was a few minutes after 1pm PST but the sky was dimmed to a soupy churning gray by the storm. What little light did make it through cast weak shadow on the sand and brittle grass. Egerton searched the shadows.

"You're being irrational," Shanwei said.

"Just check!" Egerton yelled.

"No Simon, I'm not going to check," Shanwei tried to reason with his friend. "It will take too much time and I really don't think you should be..."

Egerton's heart jumped erratically in his check. He could feel sweat ripple across his skin as the veins in his neck pumped furiously with blood. "Check!" he said again.

"No! Now come on Simon. Don't lose it on me," Shanwei coached. "You just have to keep going up the trial for a few..."

"If you don't check I'm turning around and going back to the rover," Egerton said in the depths of fear and without emotion,

"But Simon..."

"Check!"

The connection was quiet. The wind thudded into the bank. The bent and boney trees clattered against themselves. Egerton's eyes darted from the shadows to the motionless wind turbines.

"Ok, ok..." Shanwei gave in. "But it's going to take a while."

"I'll wait." Egerton stood still. The rushing blood in his neck would not let up. He was worried about his heart. Knowing he had to calm down he dropped to his hands and knees and climbed up the sand back. At the top he caught sight of the Pacific. It was so massive and violent that Egerton pushed himself deeper into the sand. It looked like a wild animal to him. Something that could kill him and wouldn't even notice.

"Simo...." Shanwei's voice broke up.

Egerton slid back down the bank, then said, "Yes?"

"Simon? Can you hear me?"

"I can hear you," Egerton answered.

"I did a one mile scan using one a Homeland Security drone," Shanwei began. "They weren't happy about giving it to me but I told them that..."

"And?"

"Simon, there's nothing on that strip of land. There are no bears, no rats, no birds...nothing," Shanwei tried to keep his voice calm. "There's nothing living out there...nothing but you."

***

The strip of sand was so narrow that Egerton could see both the bay and the ocean in his peripheral vision. The ghostly turbines were nearly behind him.

"After this thin little bit you should see a snarl of trees and another sand bank," Shanwei said. "Do you see it?"

"I see it," Egerton replied. "There's literally nothing else out here." All he could see was water and sand and clouds. "Has she moved?"

"No," Shanwei said. "Nothing's moved the whole time you've been coming close. I really think the system must have crashed."

"She knows I'm here," Egerton kept his eyes on the snarl. "If she's online at all, she knows I'm here." He noticed he was breathing much faster than before.

"Still no movement," Shanwei reported.

"Ok," Egerton forced himself to keep talking. "I'm going..."

"I'm here..." Shanwei sounded breathless as well.

As he approached the dark snarl of trees and sea grass Egerton searched for the tell-tale red of the security bots riot pads. Even in this low light he knew it should stand out. He gripped a flash light tightly in sweating hands but didn't turn it on. If the bot was online it would give him away for sure.

"Still no movement..." Shanwei said.

Egerton leaned down and crawled into the trees. The sand bank once again blocked the wind and the air clamed a little. He pushed passed a thick tangle of driftwood and dried sea weed.

"Still nothing..." Shanwei reported casually. "I really do think this think must be offline becase..."

"Oh my god..." Egerton whispered.

"What?" Shanwei grew tense. "What is it? What?"

"I see her..." Egerton went numb.

"Is she online?"

"Yes..." War Machine's system lights threw a pale glow all around the robot. The red riot pads were scratched and damaged but he could see them bright and clear. She sat on the sand with her back against the bank, slumped over so that Egerton to almost see the top of her helmet.

"It's not moving. Maybe she doesn't see you," Shanwei was hopeful.

"She knows I'm here," Egerton breathed. "She knows I'm here..."

War Machine was motionless. Egerton moved within ten feet of the bot and stopped.

"Still nothing..." Shanwei reported.

"I..." Egerton started then stopped when bot began to move.

The bot sat up slowly, almost gingerly, as if it was protecting something in her lap. Then Egerton saw what it was. Laid across her lap was the dead body of Miguel Gonzalez but that wasn't the only thing that War Machine was protecting. A small industrial robot about the size of Jimmy was doing something to the Miguel's body.

"She's moving," Shanwei panicked. "She's moving. I can see it..."

"Shhhh," Egerton whispered.

The little bot must have been from a textile factory. Its arms and fingers were designed for weaving and sewing intricate designs. Then Egerton realized what the little bot was doing.

"There's a second bot here," Egerton said softly. "It looks like it's trying to sew Miguel's body back together."

"Wha..."

Intricate seams crisscrossed the body of the dead soldier, sometimes it has sewn the uniform other times it had patched skin. Miguel's body barely looked human anymore but he was still instantly recognizable.

"It's sewing him..." Egerton said again.

Then War Machine screamed.

"What was that!" Shanwei yelled.

In a gentle but quick motion, War Machine scooped the sewing bot and Miguel's body off her lap and shot to her feet. The little bot continued its work as War Machine took a single heavy step forward. Egerton panicked.

"I'm here to take you home," he said quickly to the bot.

War Machine adjusted her stance and started to rock back and forth. Her under armor clicked and emitted a high pitched whine as the bot primed her near field defenses. There was no time for Egerton to do anything but run. War Machine attacked.

"What? What?" Shanwei yelled. "What's happening? Simon talk to me. It's moving...it's coming at you. Oh god...Simon really what's happening. "

Egerton bolted from the snarl of trees and ran toward the ocean. War Machine tore through the trees, shredding them. She screamed again as he pounded across the sand.

"Get back to the rover!" Shanwei's voice dropped into business. "Simon you have to get back to the rover. That's your only chance to..."

"I'm not going to make it!" Egerton ran, mindlessly shedding his pack or whatever gear he could.

War Machine kept coming, dropping down on all fours, she tore up the sand bank to flank Egerton and trap him between the ocean. Sand and debris flew up behind the bot. Egerton tumbled into the freezing water. The shock cleared his head for a moment and he turned. The bot ran down the bank and straight at him.

"I'm in the water now..."Egerton breathed. "I'm trapped. I can't go anywhere."

"Can you get to the rover?"

"No, I'm trapped...I'm done..."

Egerton watched as War Machine screamed and plunged into water.

"I'm done..." Egerton said again.

"No...no...no listen..." Shanwei started but Egerton was so deep in the water that he couldn't hear.

The sneaker wave surged hard and fast into the coast, knocking Egerton deep into the water. He tumbled helplessly in the tide. There was no up and no down. He was lost in the darkness and the cold. Instinctively he pulled himself into a ball as he was pushed and pulled by the water. His lungs started to burn. Blindly he pushed for the surface but found nothing but more cold water. Then another strong current grabbed him and he gave up.

***

Breaking the surface of the water, Egerton flipped on his back and gasped for air. His mouth filled with foam and the surf but his lungs got some air. He spit and coughed and breathed again.

War Machine screamed in the distance. The tide had pulled her in the opposite direction, north away from the spit and the rover. Mercifully the wave had pushed Egerton closer to Cape Look Out Point and the rover. He kicked and swam to the shore, knowing that one more big wave would slam him into the point's high rocky cliff.

Egerton stumbled and crawled out the surf. His legs gave out on him and he threw up in the wet sand. War Machine' distant scream cleared his head. He searched for the bot through the gloom of the clouds and the endless churn of the water.

When he caught sight of the bot he was overcome with an unexpected wave of compassion. She fought the tide with a relentless fury, struggling to get back on shore. But the ocean was winning.

Egerton crawled back into the rover. Jimmy and Paul were huddled at the back and didn't move when he closed the hatch.

"It's ok guys," he said still panting. "It's ok..."

Egerton fumbled with the rovers main screen and finally got it to work.

"Shanwei?" he tried. "Shanwei can you hear me?"

Nothing.

He clicked over trying a few other ports and addresses. "Shanwei? Hello? Anyone? Hello? This is Dr. Simon Egerton and I...'

"Whoa, hey man," the relaxed face of Redwood popped on the screen. "Hey! You're alive that's awesome man. Shanwei freaked out and..."

"Where is Shanwei?" Egerton asked just beginning to get his breath back.

"He's gone man. He bugged out and called the Homelanders. He's outside waiting for them..." Redwood motioned to the left of camera.

"Dr. Egerton," Jimmy said.

"One second Jimmy," Egerton held up his hand and returned to the screen. "Go get him," he said to Redwood. "Tell him I'm ok and that the security bot is in the Pacific. They're going to need a Coast Guard..."

"Dr. Egerton," Jimmy said again from the window.

"Just wait Jimmy..."

"Dr. Egerton she's coming back..." Jimmy tapped his finger on the rover's thick glass.

Egerton moved from the screen and looked out at the beach. War Machine had freed herself from the ocean and was running straight at the rover. Egerton could feel the bots violence rumbling up through the sand. There was no time. He grabbed Jimmy and Paul and jump out of the hatch.

War Machine descended on the rover and ripped it up out its deep moorings of the sand. Screaming, she tore it open and shredded the vehicle in a blind rage.

Egerton ran for the cover of the forest that covered the Cape Look Out Point. He thought maybe he could hide or at least he would be so exposed as he was on the beach.

War Machine tossed the destroyed rover aside like a broken toy and perused Egerton and the two bots.

A mound of rocks led up from the beach in into the forest. It was hard for Egerton to keep his balance as he climbed. He was careful not to hurt his bots but he was so frightened that he did feel his hand break Paul's impossibly frail arms.

"He's ok," Jimmy said.

"What?" Egerton shifted his focus to Jimmy, lost his footing and lurched into the rocks. At the last second he spun so that he landed on his shoulder and hit his head. He didn't crush Jimmy or Paul but something broke. The pain was big and broad. He worried it was his collar bone.

War Machine reached the base of the rocks and came at Egerton and the bots with no trouble.

Egerton drug himself against a tree and felt surprising calm in the few seconds that he watched the massive security race toward him. He was so calm that he didn't flinch when the Coast Guard ship dropped through the clouds and fired at War

Machine. Egerton didn't move, simply watched as the bot's attention shifted to the attacking craft.

With perfect accuracy the Guard ship took out one of War Machines legs, sending the bot sliding down the rocks and back to the sand. The next two shots removed the arms and the final did away with the other leg. War Machine was helpless to fight. She still struggled and screamed as a Guard member repelled from the ship onto the bot's chest and jammed home the kill switch.

Egerton tried to slow his breathing and check on the bots in his arms.

"Paul says that Mack wants to die," Jimmy whispered in Egerton's ear.

***

"It's like you two want to die," Director Wu said shaking her head.

"Trust me," Shanwei assured her. "I don't want to go anywhere..."

"Really," the director questioned, motioning to Shanwei and Egerton. "Look at the two of you. You're a mess." Shanwei was still confined to the robo-assit stabilization casts. Egerton had broken his left collar bone and his left arm was immobilized in a sling.

"I know it looks bad but..." Shanwei tried.

"Look," the director stopped him. "Just tell it to me again."

"We need to but Mack back into her body...a body," Egerton started. "I can't do anything with her here." He held up his personal screen. "We transferred her to my machine but if she's not in the body of a bot we're never going to learn what happened." Egerton looked tired and beaten up. Along with the sling and the bruises on side of his head, he couldn't muster much passion into his plea. He did want to find out what drove Mack crazy but it scared him and that just made him tired.

"Why do we have to put a rouge bot back into a body where it can do more damage?" the director asked.

"Watch," Egerton booted up Mack's AI. "Mack," he said to the screen.

"Yes," a calm and generic voice replied.

"Mack, how are you feeling?"

"My system is fully operational. My physical sensors are currently offline."

"Mack, what happened to Miguel Gonzalez?" Egerton asked.

The screen was silent. "Officer Miguel Gonzalez is a part of my Homeland Security Patrol Team 22-4 assigned to the American Canadian Border. I have tried to ping his suit and call his mobile but both are offline."

"Mack, what happened to Officer Gonzalez on your last patrol?" Egerton pushed.

The screen was silent. Then, "I do not have the records for that last patrol. My physical systems are currently offline and I may be experience further failures."

"She doesn't remember…"Shanwei said.

"Egerton turned to Director Wu. "Her memories are there but she can't access them. Can't or won't. I think she needs to be back in a body that's the only way we're going to get her to access the memory."

"I don't like it Shanwei," she said. "Dropping that thing back into a body after it's already taken the life of one officer seems like a bad idea."

"I get that Samantha," Shanwei replied. "But aren't you worried about the other bot? I'm worried it might happen again. You've got how many of these robotic assisted teams on the border?"

"Twenty-three," Director Wu said slowly.

"Right, that's twenty-three chances that this could happen again…"

"But I don't really understand what happened," she said.

"Neither do we…" Egerton replied. "We do know that when Mack was on patrol she was in the garment district. Officers Conley, Rodriguez and Gonzalez were following up on a tip about insurgent activity. That's when things went wrong. Mack turned on Officer Gonzalez and killed him and disappeared with the body. When I found Mack and Miguel's body that has that second bot with them…"

"This was the one sewing up Officer Gonzalez's body?" the Director said flatly.

"I think it was trying to fix him…"Egerton said uncomfortably.

"Fix? How can a textile bot fix a dead soldier?" the Director was losing her patience. She wanted this resolved quickly but her concern about the other bots on the border was clear.

"I have no idea," Egerton replied. "I think…"

"Well can't your bots ask it?" the Director pointed at Jimmy and Paul standing at the back of the office. "I was told that they can do things or talk to other robots that people can't…" she looked at Shanwei.

"Sometimes they can…" Egerton hesitated.

"Well?" the Director asked.

"Jimmy?"

"Yes, Dr. Egerton?" Jimmy answered.

"Can Paul talk to that second bot?" he asked. "Can he talk to the sewing robot we found with Mack?"

"Her name is Weaver Two," Jimmy said. "Paul says her name is Weaver Two."

"So you can talk to her?" the Director said to Jimmy.

This frightened him. He pushed himself against the wall a little.

"It's Ok Jimmy," Egerton prodded. "You can tell her. She just wants to know what happened to Mack."

"Paul says that Weaver doesn't really say much. She keeps checking her fingers for accuracy and asking for thread..." Jimmy paused. "She's a sewing bot...that's what she does."

"I don't think we're going to learn anything from the sewing bot," Egerton said. "Her AI and functions are pretty simple. She's networked out to work on a long line of other bots. The software is actually quite fascinating...it syncs all the bots on the line into a single..."

There was a quick heavy knock on the flimsy office door.

"Yes!" Director Wu called out.

Officers Conley and Rodriguez came in, their eyes scanning the room with a cold precision.

"Sorry to interrupt Director Wu," Rodriguez said efficiently. "But we heard that Dr. Egerton had just got..."

"We heard you got Mack back," Conley said enthusiastically. "Where is she?"

"Here," Egerton held up his screen.

"What?" Conley held out his hands.

"Dr. Egerton transferred your security bot to his computer so he could figure out what happened," the Director answered. "So far he hasn't had much success..."

"Why would you do that?" Conley said in a tone that bordered dangerously on insubordination. "Just ask her."

"I did," Egerton said softly.

"Mack. Mack, it's me Conley," he said, taking the screen. "Can you hear me Mack? You in there War Machine?" Then he yelled: "Hey Mack! Come on let's go!"

"What's that?" Shanwei blurted out.

"It's their rally," the Director answered a little annoyed. "It's what all the teams say before their patrol..."

"Hey Mack! Come on let's go!" Conley tried again. "You in there Mack?"

"Yes," the bot replied.

"What the..." Conley held the screen away from him as if it were diseased. "What's wrong with her? What's wrong with her voice?"

"We don't know," Egerton replied. "We think...we think..." he stopped.

"They want to load your bot into a new body," the Director said impatiently. "Dr. Egerton thinks if we do that then we'll get the answers we're looking for. But downloading into a..."

"We have an extra bot in the shop," Conley replied. "We could do it right now."

***

"We've welded in restraints to both the arms and the legs," Officer Rodriguez went through the precautions for Director Wu, Egerton and Shanwei. "For extra security we put restraints on the head, neck and chest units. The riot pads make it a little hard to secure but we got them locked down." The security bots red riot pads buckled and bulged under the straps.

"Will it hold if something goes wrong?" the Director questioned.

"To be honest Director Wu, we don't know," Rodriguez was frank. "We've never had to do this before. At least it will give us time to get out of the room."

"Almost ready," Conley said connecting the hardwire communication cable.

"You can take her right off of this," Egerton gave over his screen. "You'll find the packets..."

"I know what they look like," Conley snapped, taking the screen and returning to work.

"Where are your bots Dr. Egerton?" the Director asked.

"I left them outside," Egerton replied, glancing at Shanwei. "I wasn't sure you'd want them in the room for this."

"If they can help at all to talk to this thing," she waved her hand dismissively, "then I want them in the room."

"Ok," Egerton exited the room, glaring at Shanwei.

Shanwei smiled back with guilt.

"I'm going to start the transfer of the data first..." Conley said, turning around but stopped when he didn't see Egerton. "Where's the doctor?"

"Getting his bots..." Shanwei replied.

***

Egerton scanned the wide hallways of the shop. It was both and IT lab and a Machine shot. They did work on bots as well as drones. Jimmy and Paul were not where Egerton had left them. He panicked for a second then caught sight of them a little further down the repair bays.

"What are you doing?" Egerton asked, fussing with his sling. "Why did you..." Then he saw why.

Jimmy and Paul were looking into a repair bay watching the little Weaver bot. Her small body and delicate dexterous arms were secured to a chassis lift used to repair the underside of the big rovers and drones. She looked small and over-whelmed by the immense equipment that surrounded her. Reflexively she knitted her empty hands at irregular intervals. They clicked lightly as she looked around.

"She wants thread," Jimmy said. "I can hear her now..."

Weaver's hands clicked again.

"Come on guys," Egerton said after a moment. "I need your help with Mack."

***

Egerton returned with his bots.

"The transfer is almost done," Conley said with anticipation.

"Careful to bring her up slow," Egerton cautioned. "We want to talk to her bit by bit."

"We know the protocol," Rodriguez snapped.

"Data transfer complete," Conley reported. "Let's bring her back." He removed the hardwire and booted War Machine back to life. The body of the large bot twitched slightly then relaxed into place. She was back.

"Mack?" Conley said lightly.

"Officer Conley," Mack replied and stiffened to attention. Her voice was more animated than when she spoke from Egerton's screen. Before it had sounded like she was sedated but now she sounded awake and present in the room.

"Sheeeeee's back," Conley slapped War Machine on the arm. "Hey Mack!" he yelled. "Come on let's go!"

"Careful Officer," Director Wu warned. "This is still the same bot that took the lives of one of our men."

"Yes ma'am." Conley replied and took a step back. In the brief silence a squad of heavy drones lifted off.

"I'm going to enable her chassis," Conley said weakly.

While War Machined regained control of her arms and legs Egerton said, "Mack, my name is Dr. Egerton. I want to ask you a few questions."

"I remember you Dr. Egerton," Mack tried to shift her body to address him but was help by the restraints. Everyone in the room jumped swiftly but tried to remain calm. "I was just on your machine."

"I don't like this," the Director said and moved to the door.

"That's right Mack," he smiled at the bot, ignoring the Director. "You're held down because there's been a malfunction with your system. Do you remember it at all? Can you tell me anything about what happened on patrol with Officer Gonzalez?

Mack was silent.

"Can you remember what happened to Officer Gonzalez?" Egerton asked again.

"Why are you doing this to her?" Conley burst out. "It's not a game. Just tell her what happened and ask her why she did it."

"Relax," Rodriguez poked Conley in the back. "Relax..."

"I want Mack to pull up the data," Egerton said. "She has the memory. I want her to access it for herself so she can give us an unaltered account. The moment we tell her what happened it will alter the memory."

Conley didn't reply.

"That's why he's the expert," Shanwei said.

"Should I bring up the sensors and the network?" Conley asked.

"Yes," Egerton answered. Then said to Mack, "You were on patrol with Officers Rodriguez, Conley and Gonzalez and something went wrong. Can you remember what happened?"

"It was a routine sweep into a garment factory," Mack began to report. Her voice grew more formal. "We had swept this location multiple times in the last few weeks but never this area of the factory..."

"Ok, all the sensors and networks are live," Conley reported.

Mack's body settled into itself, as if the sensors made her more comfortable with her surroundings.

"The initial recon was that..." Mack stopped.

The room went silent then grew tense.

"Mack?" Egerton prodded.

Nothing.

"Mack?" he tried again.

Nothing.

"Come on War Machine don't bug out on us now," Conley slapped the bot on the arm again.

Then War Machine screamed. The force of the sound threw everyone backwards in the room. Jimmy and Paul fell to the floor. Shanwei tried to manipulate his robo-legs back but in the surprise forgot how to work them.

War Machine screamed again and snapped the restraints. She dove at Conley, breaking the security welds that held her arms. Conley dropped to the floor and scrambled toward the back. With a violent thrust War Machine freed her legs and attacked Officer Conley.

"No!" Rodriquez threw herself at the bot. War Machine batted her back, tossing her against the wall

"Mack! No! What are you doing?" Conley yelled. "Mack! Mack! It's me! It's me! No!"

The massive bot grabbed Conley's arm and crushed it. He screamed in pain. She pinned him to the floor and grabbed his neck. She screamed again and looked around the room.

Egerton ran out the door and down the hall.

"Where the hell..." the Director yelled after him.

Egerton knew he didn't have much time. He slipped and fell making the turn into the repair bay that held Weaver. He screamed in pain. The impact on the broken collar bone made his arms and chest feel like they were on fire. Pulling himself across the floor he grabbed and wrench from the repair tools with his good arm and bashed in Weaver's head. He slammed the metal wrench down again and again. Unable to watch, he kept his eyes closed. He struck her again and again, with wild flailing one armed swings, sometimes missing and connecting painfully with the floor but he always continued. The sound of Weaver's metal and plastic snapping sickened him.

Finally he stopped and let the wrench call to the floor of the bay. He was panting and paused to catch his breath. His entire body was in pain and his good hand was bleeding. He didn't want to look. He didn't want to open his eyes. There was no more screaming down the hall. He still didn't want to open his eyes but he did.

Weaver was destroyed. Her little head was gone. Most of her chassis was destroyed beyond comprehension. All that was left was one of her little arms still chained to the heavy drone lift.

Egerton stared at the delicate hand. He expected it to twitch or click or move but it didn't.

***

"What the hell just happened?" Director Wu asked when Egerton returned.

"You figured it out didn't you?" Shanwei said, still fussing to gain control of his robo-legs.

Egerton didn't respond.

Rodriguez was on the floor next to Conley, checking his injuries and trying to stabilize his arm.

"What's wrong with her?" Conley was in tears. He both pulled himself away from the bot and reached for her at the same time. It was as though he was stuck in a loop and couldn't stop moving. "Mack? Mack? What's wrong with her," he cried again.

"Stay still," Rodriguez pushed her partner to the floor and tried to stop him.

War Machine crouched on one knee by the wall. She held her hands close to her head.

"Will someone tell me what just happened?" the Director demanded.

Egerton held up his good hand to silence her as he approached the bot. Blood ran down his fingers and palm. "Mack? Can you hear me?" Egerton asked gently. He wiped his bloody hand on his pants. "Can you hear me Mack? Can you tell me what just happened?"

Mack turned to the sound of Egerton's voice but then saw Conley and Rodriguez on the floor. She pulled herself over to them. The sound of her riot pads squeaked on the floor. The metal and plastic in her knees joints gouged the cheap tile.

"Mack?" Conley was scared and couldn't move.

Rodriguez drew her stun gun.

Mack began to speak but stopped. She couldn't form words. She could only get out long nearly musical notes as she drew nearer to the officers.

"Shoot it!" the Director yelled.

"It won't do anything," Shanwei said, calmly. "Not against a bot."

"I think it's ok." Egerton came close to the gigantic bot as its hunched body still loomed over the officers.

Mack brought her head close to Conley's face. The staccato notes stopped and she whispered, "Conley...I...I...did it." The notes started again then stopped. "I killed Miguel."

Jimmy and Paul pulled themselves out from under the desk and Egerton glanced at them to make sure they were ok. Paul waved.

"Mack..." Conley reached for the bot.

War Machine slumped away from the two officers, going limp. She scrapped and squeaked back across the floor to the wall. She tried to pull herself upright but got stuck again on one knee.

"Hey Mack," Rodriguez called. "Hey Mack! Come on let's go!"

The bot's head flinched, almost looked over to the officers but stayed looking straight at the floor. It sounded like she was going to speak but all that came out was a low pitched whine.

***

"Let's hope these computers don't jump up and try to kill us?" Shanwei chuckled but no one laughed.

"How much longer is this going to take?" she asked.

"It's just a test..." Egerton replied but didn't look up from the two computers. He was exhausted. His broken collar bone made his entire chest hurt and he could only use one arm. As he worked his bandaged hand trembled.

"I need to wrap this up. Either you can give me a report I can file or..." the Director began.

"Look, I don't want to do this," Egerton snapped. "You want me to do it. So I'll do it but if you don't stop I'm walking out of this room...this base...this..."

"Whoa, whoa whoa," Shanwei maneuvered his legs between Egerton and the Director. "Everyone just relax. We can do this...right? No big deal right?" he looked at Egerton.

"I'll show you but you both need to shut up," Egerton growled. "I'm sorry..." he tried to explain. "I just...look...I'll be done in just a second. Let me concentrate."

"Ok," Shanwei looked to the Director. "Samantha we're all good here just keeping quiet and letting the good doctor do his work?"

The director didn't respond. She glared at Shanwei and returned to her desk.

In the silence it was clear just how thin the walls of the trailer were. Patrols went by with their bots. Small hover drones came in low and then darted about up to the border.

Shanwei scanned the office. "Where are Jimmy and Paul?" he asked forgetting the quiet he was supposed to be keeping.

Egerton didn't snap back. He just answered flatly, "I didn't want them to see this."

"Hey Scooby! Come on let's go!" a patrol yelled outside.

More drones and rovers and the sound of medication and antibiotics being pumped into Shanwei's legs.

"Ok," Egerton breathed and gingerly rubbed the bruises on the side of his head. "Come over and I'll show you."

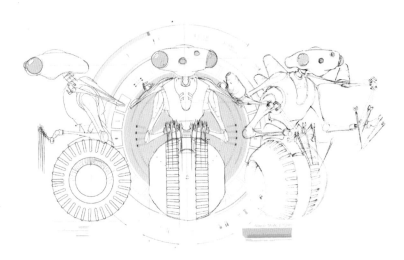

They approached and he began. "I loaded Mack and Weaver on each of these computers," Egerton gestured to the two screens with his hand. The bandages had gotten dirty as he worked on the machines. "I used the version of Mack that we had put on my machine. Before she accessed her memory files and remembered what she had done."

"Ok," the Director was listening and following intently. "I get it. You want the version before..."

"Yeah," Egerton continued. "I was also able to get into Weavers system and pull some of her system. I loaded it here." He motioned to the other computer. "Now both of these systems a closed right now. No sensors. No network. They are basically in a closed environment. Now watch..."

Egerton approached the computer that held the previous version of Mack. "Mack," he said to the screen.

"Yes," the same calm and generic voice replied as before.

"Mack, how are you feeling?" Egerton was careful to ask the same questions.

"My system is fully operational. My physical sensors are currently offline."

"Mack, what happened to Miguel Gonzalez?" Egerton asked.

The screen was silent. "Officer Miguel Gonzalez is a part of my Homeland Security Patrol Team 22-4 assigned to the American Canadian Border. I have tried to ping his suit and call his mobile but both are offline."

"Mack, what happened to Officer Gonzalez on your last patrol?" Egerton push-ed.

The screen was silent. Then, "I do not have the records for that last patrol. My physical systems are currently offline and I may be experience further failures."

"It's exactly the same," Shanwei said.

"Yeah," Egerton replied. "Just like before. Mack has no recollection."

"What does that tell us?" The director was skeptical.

"Nothing yet," Egerton grew more weary. "This is Weaver," he moved to the next computer. "She's not going to really tell us anything. "Weaver," he said to the screen. "Weaver Two? Can you hear me?" A few order numbers and protocol zipped across the screen. "She's not really made to respond. But these are codes for orders of specific thread," he pointed at the screen.

"Ok..." the director was following.

"Here's what happened," Egerton turned the computers to face each other one at a time with his right arm. He winced. All the movement was making his shoulder sore. "Up until now they were isolated, now watch." Egerton tapped on one Mack's screen then Weavers, turning on their sensors and network.

"I get it," Shanwei said.

"I don't" the Director leaned in.

"Mack," Egerton said. "Can you still hear me?"

"Yes, Dr. Egerton," War Machine replied.

"Are your sensors and network functioning..."

Before Egerton would continue Weavers screen went crazy with codes. Then War Machine started to scream through the computer speakers. Egerton didn't turn it off. He turned the volume up, filling the room with the awful sickening sound.

He watched Shanwei and Director Wu as the sound started to bother them, dig into them. Gradually they could feel War Machine's pain.

"Turn it off!" the director yelled.

Egerton didn't move. The scream continued.

"Come Simon..." Shanwei urged his friend. "We get it! Come on...

"Turn it off!" the Director lunged at the computer. She fumbled over Egerton and fell into his broken collar bone. Egerton screamed in pain, trying to shake it off.

"Simon!" Shanwei yelled.

Egerton pulled himself back to the desk and silenced the computer.

"What the hell do you think you're doing?" the Director asked from the floor, she struggled to retain her dignity and authority.

"That's what it sounds like when a bot goes insane," Egerton turned his back on the pair. He tapped on the screens, transferring the files to the Homeland Security Servers then deleted them from the machines.

"That's crazy," the Director stood and moved back behind her desk.

"It might sound crazy but it's not," Egerton replied with purpose. He packed up the computers and gathered his things. "It was the network connected that linked them. When Mack entered that factory and connected with Weaver something happened. I think Mack accidentally damaged Weaver One, the bot that sat just in front of Waver Two. The weaver bots work in a tight configuration.

"So?" the Director wasn't following.

"So, with Weaver One destroyed I think Weaver Two latched onto War Machine, somehow got into her brain."

"And that did it!" Shanwei called out. He understood.

"What are you two talking about?" Director Wu was still lost.

"When Weaver latched on to Mack there was nothing War Machine could do," Egerton explained. "Weaver was in her brain, making the incessant call for thread. Over and over. Both Weaver and Mack malfunctioned and pushed each other further. It drove Mack insane..."

"You can't drive a bot insane," the Director said.

"Yes you can," Egerton replied. "Just like you can drive a human insane. There was nothing Mack could do. She was doomed. All she could do was attack to try and stay alive...and then she killed on e of the people she cared for most in the world."

"And that just made it worse," Shanwei added.

"Yeah," Egerton was tired. "It drove Mack into the rage we just saw, making her turn on the people she loved and swore to protect. She killed Miguel and then it just got worse..."

"So when the AI's mingled something happened?" Shanwei was processing. "Like just connected over a network?"

"After killing Miguel, Mack got worse. She'd killed one of the three people she loved and cared for most and couldn't think of anything else to do but flee. She was bonded to Weaver and couldn't bear to leave Miguel behind..." he paused. "I don't know maybe it was shame or maybe she thought that weaver could put Miguel back together..." Egerton massaged the skin under the sling.

"How am I supposed to explain security bots that go crazy when they get on the same network as other bots?" the Director asked. "How do I make sure this doesn't happen with my 23 other Patrols?"

Egerton paused and looked Director Wu directly in the eye and replied, "You don't."

"What do you mean I don't?" she spat back. "What am I supposed to do? What the hell happened?"

Egerton moved to the door. "I don't know what happened. I don't think we'll ever know. I don't think we can know..." he took a breath. "...and I don't think we can stop it from happening again."

"But..."

Egerton walked out the door.

***

"Wait!" Shanwei chased after Egerton, working his robo-legs. They weren't meant to run but he was pushing them as fast as they could go. "Wait Simon..."

"No," Egerton replied, not stopping. He walked out of the field office trailer and found Jimmy and Paul waiting for him. "I'm not stopping. I'm not waiting and I'm not coming back!" Egerton turned.

"What? Why?" Shanwei grew concerned.

"I'm done," Egerton said flatly.

"What?"

"I'm done. No more," Egerton couldn't look at his partner. "I can't do this anymore. I kept it together to finish this job...I did what I said I'd do...I don't...I don't want to get you in trouble but..." Egerton cracked. His breath came in deep gasps and he started to tear up. "I can't do this..." Jimmy and Paul moved in closer to his legs.

"What? What's wrong? You mean this," Shanwei pointed to his legs and Egerton's injuries. "This is nothing. Come on man you know we'll be fine in a week or so. No big deal."

"It's not that..." Egerton shook his head. "I don't care about that..."

"Then what? What is it? Tell me and I'll fix it."

"You can't fix it..." Egerton snapped. "Look this was great...I mean when we started all this work it was really great and interesting but I can't do it anymore. It's too much. You don't know what it does to me. I can't do *this* again..."

"What is it?" Shanwei looked around desperately but there was no one to help. The drones passed and the patrols walked by the trailer. They didn't know who these two men with the two bots were and they didn't care.

"I drove a bot insane *twice* today," Egerton snapped into a static and calm voice. "You don't know what that means... "

"What?" Shanwei asked. "What does it mean?"

"You don't know what this does to me. All of this...all of these bots are so troubled. I can't take it anymore. I can't do it anymore." Egerton stopped. "I don't think you know what all this means..."

"What?" Shanwei asked again. "What does it mean?"

He looked down at Jimmy and Paul then scanned Homeland Security's Border operation. "It means I'm done," he answered then took Jimmy and Paul and walked away.

Shanwei wanted to call after him but he didn't know what to say.

Gear Up for Action

Sneak Peek! This is a summary of the chapter to come. Below is the journey we are taking into the future of robots.

Your 21st Century Robot is almost ready to go! The final step in your build is to connect all the systems and wake up your robot. USC's Ross Mead and the Olin crew walk us through the final steps and show you how to share your designs with the rest of the world.

Then Maya and Simon show us how to build and share apps for your robot. Iteration and sharing is essential to our Manifesto. When our collaborators share their visions for the future of robots we can truly build amazing things!